S0-BXL-538

STUDIES IN

THE LATEGLACIAL OF NORTH-WEST EUROPE

Related Pergamon Titles of Interest

ANDERSON: The Structure of Western Europe

BOWEN: Quaternary Geology (A Stratigraphic Framework for Multi-disciplinary Work)

GERASIMOV: International Geography - 76 (Proceedings of the 23rd International Geographical Congress, Moscow) 12 Volume Set
Volume 1: Geomorphology and Palaeogeography

GRAY & LOWE (Editors): Studies in the Scottish Lateglacial Environment

OWEN: The Geological Evolution of the British Isles

*SEARS: The Quaternary History of the Ocean Basins

These books are available for free inspection for a period of 60 days (except where marked *) under the Pergamon textbook inspection copy service on application to your nearest Pergamon office.

STUDIES IN THE LATEGLACIAL OF NORTH-WEST EUROPE

Including Papers Presented at a Symposium of the Quaternary Research Association Held at University College London, January 1979

Edited by

J. J. LOWE
City of London Polytechnic

J. M. GRAY
Queen Mary College, London

J. E. ROBINSON
University College London

PERGAMON PRESS

OXFORD · NEW YORK · TORONTO · SYDNEY · PARIS · FRANKFURT

U.K. Pergamon Press Ltd., Headington Hill Hall, Oxford OX3 0BW, England

U.S.A. Pergamon Press Inc., Maxwell House, Fairview Park, Elmsford, New York 10523, U.S.A.

CANADA Pergamon of Canada, Suite 104, 150 Consumers Road, Willowdale, Ontario M2J 1P9, Canada

AUSTRALIA Pergamon Press (Aust.) Pty. Ltd., P.O. Box 544, Potts Point, N.S.W. 2011, Australia

FRANCE Pergamon Press SARL, 24 rue des Ecoles, 75240 Paris, Cedex 05, France

FEDERAL REPUBLIC OF GERMANY Pergamon Press GmbH, 6242 Kronberg-Taunus, Pferdstrasse 1, Federal Republic of Germany

First edition 1980

British Library Cataloguing in Publication Data

Studies in the lateglacial of north-west Europe.
1. Geology, Stratigraphic - Pleistocene
2. Geology - Europe
I. Lowe, J J II. Gray, J M III. Robinson, J E
551.7'92'094 QE697 79-42965

ISBN 0 08 024001 1

In order to make this volume available as economically and as rapidly as possible the authors' typescripts have been reproduced in their original forms. This method has its typographical limitations but it is hoped that they in no way distract the reader.

Printed and bound in Great Britain by William Clowes (Beccles) Limited, Beccles and London

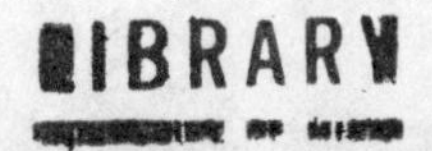

Contents

Contributors

Dr. C.J. CASELDINE
Department of Geography, University of Exeter, Rennes Drive, Exeter, EX4 4RJ

Dr. G.R. COOPE
Department of Geological Sciences, University of Birmingham, PO Box 363, Birmingham, B15 2TT

Dr. A.G. DAWSON
Department of Geography, Lanchester Polytechnic, Priory Street, Coventry, CV1 5FB

Dr. J.M. GRAY
Department of Geography, Queen Mary College, University of London, Mile End Road, London E1 4NS

Dr. M.J. JOACHIM
Department of Geological Sciences, University of Birmingham, PO Box 363, Birmingham, B15 2TT

Dr. A.R. LORD
Postgraduate Unit of Micropalaeontology (Geology), University College London, University of London, Gower Street, London WC1E 6BT

Dr. J.J. LOWE
Geography Section, City of London Polytechnic, Calcutta House, Old Castle Street, London E1 7NT

Dr. J.B. MACPHERSON
Department of Geography, Memorial University of Newfoundland, St. John's, Newfoundland, Canada A1B 3X9

Dr. J. MANGERUD
Department of Geology, Quaternary Institute, University of Bergen, Allégt. 41, N-5014 Bergen, Norway

Dr. P.D. MOORE
Department of Plant Sciences, King's College, University of London, 68 Half Moon Lane, London SE24 9JF

Dr. J.E. ROBINSON
Department of Geology, University College London, University of London, Gower Street, London WC1E 6BT

Dr. J.B. SISSONS
Department of Geography, University of Edinburgh, High School Yards, Edinburgh EH1 1NR

Mr. D.G. SUTHERLAND
Department of Geography, University of Edinburgh, High School Yards, Edinburgh EH1 1NR

Dr. M.J.C. WALKER
Department of Geography, St. David's University College, Lampeter, Dyfed SA48 7ED, Wales

Professor W.A. WATTS
University of Dublin, School of Botany, Trinity College, Dublin 2, Eire.

Acknowledgements

We would like to thank all those who so readily offered help or advice in the production of this volume, especially those who agreed to act as referees for the papers, often at extremely short notice. This volume has been assembled thanks to the help of numerous friends.

Several colleagues also helped with the planning and organisation of the Quaternary Research Association meeting held in January, 1979, where many of the papers here included were first presented. Special thanks are due to John Catt, Jim Rose and Peter Worsley, to those who took the Chair for the various sessions, to all the contributors at the meeting, and to the research students (David Bridgland, John Ince and Stephen Lowe) who ensured that the meeting ran smoothly.

We acknowledge with thanks cartographic and photographic assistance from Don Shewan, Penny Tommis, Mavis Teed and Patrick Foster (City of London Polytechnic), Ray Crundwell and Peter Newman (Queen Mary College). Advice, assistance and encouraging remarks were always available from Peter Henn and other staff at Pergamon Press. Jeanette Lowe helped with the proof-reading, and Liz Ellis typed and corrected the final camera-ready copies with amazing speed and skill.

To everyone involved - our sincere thanks.

John Lowe

Murray Gray

Eric Robinson

London, August, 1979

Introduction

Workers in other parts of the world must be forgiven for thinking that we Europeans seem to be obsessed with the period between about 14,000 and 10,000 BP. Although many parts of the world experienced climatic changes at this time with concomitant environmental responses that are of considerable interest, the European Lateglacial has particular characteristics that make it unique in both space and time.

Present knowledge suggests that the dramatic environmental changes taking place in Europe at this time were related to changes in oceanic circulation and in this respect western Europe is in a particularly sensitive location since its climate is greatly influenced by the North Atlantic Drift. Studies of ocean floor sediments by CLIMAP scientists indicate that polar waters retreated northwestwards at the beginning of the Lateglacial only to return again towards its close, during the Younger Dryas. Thus WATTS suggests that the beginning of the Lateglacial marks "a false start" to the present interglacial. In reviewing regional variations in vegetational responses across Europe he notes that the Younger Dryas was most strongly felt in NW Europe but diminished in its impact towards the east and south. Similarly N America apparently experienced nothing comparable, probably due to the fact that the oceanic circulation changes at this time were mainly affecting NW Europe. Thus the dramatically irregular transition from the last glaciation to the beginning of the present interglacial appears to have been a predominantly W European phenomenon. As WATTS remarks, it is usual for major climatic events to have a world-wide expression, but here is one that appears to have affected only one part of the world.

The climatic variations taking place during the Lateglacial brought about numerous environmental responses. One of these was in the ice masses and MANGERUD reviews ice-front variations of the Scandinavian ice-sheet during the Lateglacial, and relates regional differences to climatic and topographic factors. While the Younger Dryas brought a significant readvance of the Scandinavian ice-sheet, in Britain most or all the resultant glaciers and ice-caps developed anew. SISSONS describes the palaeoclimatic inferences that may be made from the reconstructed glaciers mapped as existing in N Britain at this time. DAWSON describes the effects of the cold climate on the coast of W Scotland where rapid erosion of rock appears to have occurred during the Younger Dryas. Marked responses were also felt in the distribution of plants (CASELDINE, MACPHERSON, MOORE), beetles (COOPE and JOACHIM) and marine microfauna (LORD, ROBINSON) during the Lateglacial. In all, the landscape changes at this time were considerable and they occurred with a rapidity that is seemingly unparalleled in the geological record.

The availability of radiocarbon dating for this period, and the rich, well-preserved biological, sedimentological and geomorphological evidence provides us with a unique stratigraphic record and environmental history, but as the papers in this book illustrate, unique stratigraphic problems are also encountered. In particular, problems of radiocarbon dating in the Lateglacial are discussed by LOWE and WALKER, and by SUTHERLAND, while the final paper by LOWE and GRAY discusses the basis of the stratigraphic subdivision of the Lateglacial of NW Europe. This paper encapsulates two main themes introduced in preceding papers; first, that stratigraphic schemes are essentially a synopsis of biological, lithological or morphological parameters, but evaluations of these parameters, and hence their palaeoenvironmental implications are often debatable; and secondly, absolute chronology and correlation within the Lateglacial rest heavily on radiocarbon dating, but in the last few years new problems in radiocarbon dating Lateglacial materials and in interpreting the dates have come to light.

For consistency, throughout the book, in papers concerned with continental Europe, we have tried to use the terminology and definitions that are clearly set out in Mangerud *et al.* (1974) with subsequent amendments in Mangerud and Berglund (1978), and for those concerned with the British Isles we have used terms of Mitchell *et al.* (1973) and those defined by Gray and Lowe (1977) for Scotland, *i.e.*

Lateglacial Interstadial: the period between the start of the apparently rapid thermal improvement that occurred between about 14,000 and 13,000 BP and the beginning of the marked thermal decline that took place around 11,000 BP.

Loch Lomond Stadial: the period between the start of the marked thermal decline that occurred around 11,000 BP and the international chronostratigraphic boundary for the beginning of the Flandrian (Holocene), *viz.* 10,000 BP (roughly equivalent to the Younger Dryas of continental stratigraphy).

It must be emphasised however that our definitions may not be totally accepted by all the authors contributing to the volume, nor strictly adhered to in the text. Ideas on Lateglacial stratigraphic subdivision vary throughout Europe in terms of the number, duration and dates of subdivisions that are recognised as indicating major environmental modifications, and environmental changes are by nature time-transgressive. However, it will be shown that there appears to be a measure of uniformity in the records for NW Europe, and a suggestion for a general stratigraphic scheme, applicable to NW Europe as a whole is presented in the final paper. This requires certain modifications to terms defined above.

This volume grew out of a symposium of the Quaternary Research Association on "The Lateglacial environment of the British Isles and possible correlations with NW Europe" held at University College London in January, 1979. Many of the papers here included were first presented at this symposium.

The book sets out to illustrate the nature of some of the environmental changes that took place and the efficacy of landscape modification during the Lateglacial, and discusses the basis for subdividing the Lateglacial period, the extent to which an absolute chronology can be erected by radiocarbon dating, and the confidence which can be attached to correlations of Lateglacial events within NW Europe. If the volume provokes deep discussion of the important problems associated with this fascinating period of geological history, it will have achieved its aim.

Regional Variation in the Response of Vegetation to Lateglacial Climatic Events in Europe

W. A. Watts

(University of Dublin)

ABSTRACT

There is great regional variation in the intensity of response of vegetation to Lateglacial climatic events in Europe. Between 11,000 and 10,000 radiocarbon years BP, a severe climatic deterioration brought about major changes in the species composition of vegetation in Ireland and Britain. In continental NW Europe most of the same species persisted from the preceding period and changes in vegetation were largely quantitative, although heaths developed over sandy soils. In the Alps there was little or no change in vegetation at many sites, although there was inflow of inorganic materials into some lake basins. There is weaker evidence for a climatic deterioration which caused upland erosion between 12,000 and 11,800 BP. This can be seen in evidence from Ireland and Britain but it is weakly expressed and in evidence from only some sites in Denmark and northern Germany. There is no evidence for such a deterioration in the Alps. Very tentative climatic curves are presented to draw attention to regional differences which have not been sufficiently appreciated in the literature.

The deterioration from 11,000 to 10,000 BP is correlated with a readvance of polar water into the eastern North Atlantic. Evidently it affected coastal areas strongly, but had a progressively diminished effect away from the ocean. Lateglacial climatic and vegetational fluctuations claimed elsewhere are regarded as inadequately founded at present. The apparent uniqueness of the Lateglacial climatic events is noted. It is questionable whether the Lateglacial can be regarded as an interstadial or complex of interstadials. The vegetation development seems to have the character of an abruptly interrupted interglacial cycle. The chronozone concept is not found to be preferable to direct correlation of profiles by radiocarbon dates.

INTRODUCTION

The literature on the Lateglacial period is now vast, and it is not possible to review it fully within the space limitations of a short article. This account is mainly concerned with a consideration of the vegetational and inferred climatic record of the Lateglacial on a continental scale, taking account of differences as well as similarities among regions. The intention is to demonstrate that, in spite of the relatively superficial similarities that provide a basis for stratigraphic correlation, there are significant differences among regions in the details of the climatic record and in the intensity of vegetational response to climatic change. Also discussed are geographical areas affected by the Lateglacial climatic episode, its unique character and cause, and its claims to be regarded as an interstadial.

'Lateglacial' is used here as an informal term to describe the vegetation and inferred climate of the Late Pleistocene interval between approximately 14,000 and 10,000 years ago. It is followed by the Flandrian (or 'Holocene'), from approximately 10,000 years ago to the present. Sites referred to in the text and others at which recent studies have been carried out are located on Figs. 1-4. For location of areas outlined in Figs. 1-3 see Fig. 4. Dates are cited in radiocarbon years before present (BP). Original papers should be consulted for standard errors and laboratory numbers, but dates with large errors are not cited.

Commenting on the use of pollen zones of Jessen (1938) to describe vegetational changes in the Lateglacial and Flandrian of Denmark, Mangerud *et al.* (1974) observe that Jessen's zone system is "not used exclusively as a biostratigraphical classification, but rather as a combination of biostratigraphical, climatostratigraphical, and chronostratigraphical classifications. For the present-day need of precision, this is not satisfactory" (p.111). The same criticism is valid for the use that has been made of pollen zones to interpret the information about climate and vegetation contained in Lateglacial pollen diagrams from western, northern, and Alpine Europe. The zone system of Iversen's classic paper (1954) on the Lateglacial of Denmark has been very widely used, and chronological assumptions have become attached to the system (*e.g.* Wegmüller and Welten, 1973). The expectation that the same series of oscillations and the same chronology should be found everywhere is not well founded in principle, as, indeed, was already discussed by Faegri and Iversen (1964, Fig. 18 and pp.110-111). Some principles involved in the interpretation of Lateglacial pollen diagrams are discussed below.

THE INTERPRETATION OF THE FOSSIL RECORD

A pollen diagram is a record of the growth, stability, decline, or extinction of plant populations in one locality (Watts, 1973). This is particularly well seen where pollen data are presented as influx diagrams (Davis, 1969), which highlight the independent behaviour of the taxa. Each species behaves independently of others because its response to climatic or other environmental factors is determined by its own genetic inheritance, which may itself be sufficiently variable throughout its geographic range to allow for some small-scale diversity in local response. Independence of behaviour is limited by competition with other species for suitable habitat, so that the capacity of a population for expansion or contraction in response to climatic change may be determined by the assemblage of competitors present at each locality (see discussion in Davis, 1978). These considerations suggest that each pollen profile must be studied and dated as a local record in its own right, because each may differ from neighbouring sites in environmental conditions, competition, and timing of population changes.

Migration and Lag Effects

Species expand their populations by migration in response to climatic and ecological opportunity. Speed of migration has natural limits imposed by the ability of plants to produce, disperse, and establish propagules, and by the age at which fruiting can first take place. Migration may be delayed in a favourable climatic or competitive environment if time is required for suitable habitats to become available for colonisation, for example, for the build-up of organic material and populations of micro-organisms in recently deglaciated raw soils. Vegetation may be 'out of phase' with the climate because these lag effects have delayed the arrival of migrants. Migration is inherently time-transgressive, although rates of migration may be too fast for differences in time between closely spaced sites to be distinguished by radiocarbon dating.

It may not be possible to discriminate between a pollen peak caused by a population invading from another region and one caused by local expansion of a species previously present but climatically little favoured. A good example is provided by Belle Lake, Ireland (Craig, 1978), where juniper survived the cold period between 11,000 and 10,000 BP with low frequencies and expanded greatly in the early Flandrian. At other Irish sites, such as Ballybetagh (Watts, 1977), juniper disappeared temporarily from the record. In this case the Flandrian migration had only a short distance to cover, and expansion may have taken place rapidly over a short period.

Range Margins and Ecotones

The response of a plant population to climatic change is likely to be greatest near the margin of its range (tolerance). A relatively small change in mean July temperatures may not result in a measurable response from a species near the middle of its range, but it may cause notable expansion or contraction at the margin, with lesser effects of briefer duration in intermediate localities. There is a well-known example in the invasion of forest by prairie along the prairie/forest border in Minnesota during the Flandrian (McAndrews, 1967). The contemporary forest farther east in the United States shows no clear record of the dry phase (Wright, 1971). The recognition of the location of range margins or ecotones between plant formations in the past is important to an understanding of climatic history, because it is in the ecotones that change is most clearly expressed.

Zonation of Pollen Diagrams

The criticism of Mangerud *et al.* (1974) that the older style of pollen zone of Jessen (1938) and Iversen (1954) is unsatisfactory because it confuses biostratigraphical, climatological, and chronological criteria is accepted here. In fact, because the events recorded in pollen diagrams may be both time-transgressive and variable in expression from site to site, it may be doubted whether zonation should be used at all (*cf.* discussion in Watts, 1977, pp.275-277), except as local zones to assist in the description of events at each site. Local pollen sequences are the primary information of pollen analysis. A local profile may have much in common with those of neighbouring sites, but it also retains the record of local variation in the vegetation cover in the past, and of its relationship to migration, competition, soil, altitude, exposure, climate, and other environmental considerations. Radiocarbon dates are the best basis for comparison of profiles, because the development of vegetation at two sites being compared has a known chronology, and factors such as time-transgressiveness of migrating populations (Davis, 1976; Watts, 1979) can be allowed for. Mangerud *et al.* (1974) may have these considerations in mind when they recommend the use of chronozones. The concept may be superfluous, however, if a comparison between profiles is based solely on time, expressed as radiocarbon dates. No additional information is provided, for example, by calling a profile segment already dated to between 11,800 and 11,000 BP the Alleröd Chronozone. It has the disadvantage of implying that the chronozone is of homogeneous character and duration over wide areas, *i.e.* re-introducing the very confusion it was intended to expel. Probably the most practical way to compare pollen diagrams is to use pollen assemblage zones (Cushing, 1967; Birks, 1973; Berglund, 1976) dated by radiocarbon and presented as time-space diagrams. Even here, the assemblages tend to be weakly defined in terms of the vegetation they represent and the climatic parameters that can be inferred, and the beginning and ends of zones are inherently difficult or impossible to define, because they represent vegetational transitions that themselves occupy time.

Plant Communities and Formations

Plant communities or associations in the sense of plant sociologists are difficult to recognise in the Lateglacial partly because the periods of stable climate during which migration and invasion were relatively static may have been very short. The species assemblages, especially those dominated by herbs, appear to lack analogues in the modern flora. For this reason pollen surface sampling cannot be used to interpret Lateglacial assemblages very readily, and it is not easy to use transfer functions (Webb and Bryson, 1972) to infer the climate (*cf.* discussion in Davis, 1978). Because of the difficulty of reconstructing plant communities and their instability, the reconstruction of climate is likely to depend more on what is known of the autecology of important species. This may leave climatic values unquantified and in an unsatisfactory degree of vagueness, but it may correspond to the true level of knowledge.

Timing

Time may itself be difficult to measure because of the inherent difficulty of dating sediments that may be relatively inorganic and that may have accumulated extremely slowly. Radiocarbon dates may be carried out on core-segments that themselves represent some hundreds of years of sedimentation (*cf.* Cam Loch, Pennington (1975a) where radiocarbon dates are based on seven contiguous samples that represent virtually the complete Lateglacial core; each segment may represent up to 400 years). This results in loss of resolution in dating events (see also discussion in Gray and Lowe, 1977). It may be unavoidable at many sites, but it makes for difficulty in exact timing of local events and in correlation. Other sources of dating difficulty include 'old' carbon in carbonate-rich sediments as well as secondary redeposition of fragments of coal, lignite, or other carbon-bearing materials (see also Sutherland, this volume; Lowe and Walker, this volume).

Summary

These various considerations suggest that pollen diagrams with a local zonation, correlated with others by radiocarbon dates, provide the best data base for a continental view of variation in the vegetation and climate of the Lateglacial. The climate is inferred largely from autecology of important plant species. It should be possible to identify the number of climatic events within the Lateglacial and their broad climatic character, but it may not be possible to date their initiation accurately because of lag effects and the inherent difficulty of obtaining high resolution with radiocarbon dates. The strongest expression of climatic events is in ecotonal areas, which are important to identify and locate. Some very clear quantifiable evidence of climatic change comes from non-botanical sources. For example, evidence comes from the record of cirque glaciers between 11,000 and 10,000 BP (Pennington, 1977a; Colhoun and Synge, 1979), aeolian phenomena such as the coversands of the Netherlands and adjoining regions, lithostratigraphic evidence for erosion and solifluction (Craig, 1978), and the foraminiferal evidence of change in ocean surface temperatures (Ruddiman *et al.*, 1977). The most detailed information, however, relates to the use of reconstructed firn-line altitudes for Loch Lomond Stadial glaciers. Climatic interpretations based on these data are discussed by Sissons (this volume) for the information available so far from the British Isles.

Lateglacial pollen diagrams from Europe are now reviewed against this background, paying special attention to detailed diagrams, mostly published since 1960, that have counts for *Juniperus* (overlooked in earlier work) and that ideally are supported by several radiocarbon dates.

REGIONAL SEQUENCES

Southern Scandinavia and Continental NW Europe (Fig. 1)

Fig. 1 Lateglacial sites in Denmark and surrounding regions

The classic analysis of the climate and history of vegetation in the Lateglacial is the study by Iversen (1954) at Böllingsö in Denmark. Iversen identified two periods of climate warming, the Bölling and Alleröd oscillations, separated by cold episodes. Iversen's essential conclusions and interpretation are presented in Table 1.

Unfortunately, the radiocarbon dates from Böllingsö are unrelated to Iversen's profile, which remains undated. Dates cited for the Bölling oscillation derive from

Pollen Zones - Denmark			Chronozones - Scandinavia (C14 yrs. BP)	
			Preboreal cz.	10,000 - 9,000
Zone	III	Younger Dryas Period	Younger Dryas cz.	11,000 - 10,000
Zone	II	Alleröd Period	Alleröd cz.	11,800 - 11,000
Zone	Ic	Older Dryas Period	Older Dryas cz.	12,000 - 11,800
Zone	Ib	Bölling Period	Bölling cz.	13,000 - 12,000
Zone	Ia	Daniglacial Tundra Period	?	? - 13,000

TABLE 1 Lateglacial Pollen Zones (Iversen, 1954) and Chronozones (Mangerud *et al.*, 1974)

Gaterslebener See (Müller, 1953; Firbas *et al.*, 1955) and from Usselo (Tauber, 1960). *Juniperus* was not recorded by Iversen, but a new study by Stockmarr (1975) suggests that it may not have been very important at Böllingsö. Stockmarr's work was carried out on peat exposed in the Böllingsö Canal, 1½ km from the former lake studied by Iversen. His diagram (Böllingsö, Canal 2) is very similar to Iversen's especially in the curves for *Betula*, Cyperaceae, and *Hippophaë* and it is at least arguable that it covers virtually the same period of time. There is a basal date of 11,970 BP. However, Stockmarr believes that his diagram begins in Zone Ic (*cf.* Table 1) and that at least some of the numerous dates reported in his study are unreliable because of disturbance and secondary redeposition of sediment. Iversen's diagram shows very considerable rebedding of pollen in Zone Ia (to 75%) and Ic (over 80%). The amount of secondary material present suggests that it may never be possible to date Böllingsö reliably. Thus the site of Böllingsö is unsatisfactory as a type-locality for the Bölling oscillation.

To what extent is the Böllingsö sequence repeated elsewhere? SW Norway (Mangerud, 1970) and Sweden south of Göteborg (Berglund, 1976) were ice-free before 12,500 BP, so in principle it should be possible to find a sequence like that at Böllingsö throughout Denmark and deglaciated areas of Norway and Sweden, except for coastal areas invaded by the *Zirphaea* Sea (Krog and Tauber, 1974).

At Ruds Vedby (Krog, 1954) the Lateglacial sequence shows high levels of herb pollen and rebedded pollen bracketing the Alleröd phase of high values for *Betula* (to 60%). The middle of the Alleröd phase has over 10% of *Juniperus* pollen. A Bölling oscillation below cannot be identified.

The deposit of Flådet (Fredskild, 1975) is also rather undifferentiated and lacks dates. Fredskild believes that sedimentation began in Zone Ic. The Younger Dryas Period (Zone III) is well marked at Flådet by increases in *Artemisia*, *Salix herbacea* type, rebedded pollen, and content of sand.

At Vallensgaard Mose, Bornholm, Iversen (1954) did not subdivide Zone I. At Håkulls mosse (Berglund, 1971, 1976) Zone Ic is identified by an increase in *Dryas*, *Artemisia* and Chenopodiaceae at the expense of *Betula*. The base of the profile is dated to 13,020 BP with 12,660 BP for the "Bölling" level. Åkerhultagöl (Berglund, 1976) shows a very strongly defined Younger Dryas Period, with a marked reduction in tree pollen dated to between 11,200 and 10,400 BP.

At Brøndmyra in western Norway, Chanda (1965) has shown that *Salix* and *Betula* invaded a herb-dominated landscape after 13,000 BP. By 12,650 BP *Betula* was providing 70 - 80% of the pollen. Subsequently, a long slow decline in birch percentages accompanied by increases in *Artemisia, Rumex,* Chenopodiaceae, and Caryophyllaceae took place from 11,300 to 10,800 BP, followed by increases in *Betula* and *Pinus* after 9,000 BP. By analogy with Britain and Ireland, the *Artemisia*/Caryophyllaceae phase may have most claims to be "Younger Dryas", and it is notable that this section of Chanda's diagram has sand, suggestive of soil erosion, which is not present above or below. No Zone Ic can be identified.

On the small island of Blomöy, Mangerud (1970) has shown a sequence of high grass percentages with herbs, then a birch/willow assemblage with over 60% tree and shrub pollen, followed by an increase in sedges and decline in birch before the Flandrian birch rise begins. The grass peak zone is dated to 12,070 BP and the birch decline to 10,940 BP. The Blomöy bog overlies till which in turn overlies littoral sediments with wood, bone, and shells. Dates of 12,200 and 12,700 BP come from wood and shell respectively. The Blomöy data suggest a local ice readvance during Zone Ic. Data from near Bergen provide evidence for local readvances in both Zones Ic and III (Mangerud, 1970).

In North Germany at Glüsing (Menke, 1968), a long pollen diagram shows little differentiation into phases, although Menke identified both the Bölling and Alleröd oscillations and even a new Grömitz oscillation and Meiendorf interval below the Bölling, correlated with the Susaca interstadial of van der Hammen and Vogel (1966)! After a period with high grass and sedge pollen with values for pine and birch generally below 10%, birch increased to 20% and *Helianthemum* and *Hippophaë* were prominent. Subsequently birch rose further to 40-50%, mainly at the expense of grass. *Empetrum* occurs in the upper part of the diagram, especially near the top, where the Younger Dryas period is identified. The evidence for oscillations is very slight, and the whole diagram can be interpreted as a slow step by step increase of birch forest cover in what was at first an open landscape, with development of *Empetrum* heath in the Younger Dryas.

At Rabensbergmoor (Usinger, 1975) there is a similar little-differentiated diagram showing the immigration of birch. Usinger's other site, Kubitzbergmoor, shows three very clear zones. An early zone rich in *Hippophaë* and *Betula nana* type pollen is succeeded by predominant tree-birch (Alleröd). Subsequently dwarf birch, juniper, and *Empetrum* expand, presumably in the Younger Dryas, and tree-birch declines. A hiatus is suggested at the end of the tree-birch zone, but the evidence is not compelling. A Bölling oscillation cannot be identified.

Lesemann (1969) shows a similar sequence at Siemen (Profile II), a more southerly site where pine was more important than birch towards the end of the Lateglacial.

At Westrhauderfehn, Behre (1967) records an early pioneer assemblage with juniper, *Hippophaë, Thalictrum, Artemisia,* and *Helianthemum.* A birch peak with over 70% of the pollen sum follows. Behre attempts to identify Iversen's zones Ib and Ic, apparently on the evidence of a small fluctuation in the *Betula* curve, but the evidence is slight and unconvincing. The Younger Dryas in this sandy region shows expansion of heath plants and reduction in pine and birch. Pine reached 30% of the pollen sum at this site in the Lateglacial but did not exceed birch.

In contrast, in the otherwise similar Waskemeer site in the Netherlands (Casparie and van Zeist, 1960), pine rose to 50% and briefly exceeded birch near the end of the Lateglacial.

High pine values with Laacher See volcanic ash are known from Rotes Moor in the Rhön district (Beug, 1957), but the record is difficult to interpret because of very high sedge values and lack of a juniper record. The pine limit must have extended from the Netherlands to the lowlands of N Germany in the Lateglacial without reaching

Schleswig-Holstein or Denmark (see Firbas, 1950).

All the sites mentioned above have slight or no evidence for a Bölling interstadial. Let us now consider sites that appear to show two well-differentiated warm phases in the Lateglacial.

At Usselo (van der Hammen, 1951, Profile A), a sequence with basal *Hippophaë*, high herbs, and low trees is referred to Zone Ia, a later birch peak with up to 45% of the pollen is identified as Ib, and there is a final Zone Ic with lower values of birch, less than 20% in one sample. The birch fall is caused by a rise in Cyperaceae only; no other taxon shows clearly correlated changes. *Plantago, Helianthemum,* and Caryophyllaceae occur throughout Zone I but decline in II, where birch exceeds 50% and pine has 20 to 30% of the pollen throughout. In Zone III, *Empetrum* records the development of heath-like vegetation on coversands as trees fall back. Usselo (Profile B, but not Profile A) has been dated by Tauber (1960).

Profiles A1 and A6 from a former lake, Gaterslebener See (Müller, 1953), show well-characterised pollen zones. Laacher See volcanic ash defines the Alleröd level. It occurs shortly after pine pollen became more frequent than birch in a forested period, as is true at many sites from western Switzerland to the Berlin area (Müller, 1965; Wegmüller and Welten, 1973). In the Younger Dryas interval, *Artemisia*, grasses, and sedges expanded as pine declined, but even then, in this relatively southern site, *Pinus* still provided up to 50% of the pollen. The base of the most detailed profile (A6) has largely sedge pollen with *Betula nana* and *Hippophaë*, with a date of 13,250 BP on birch and willow stems which were up to 5 cm thick. Subsequently, birch and willow expanded, followed by pine to 30% and total tree pollen to 70%. This level, considered to belong in the Bölling oscillation, has been dated to between 12,700 and 12,300 BP (Firbas *et al.*, 1955). A long Zone Ic (Profile A6) has up to 50% tree pollen with fluctuating values of birch, willow, and pine and relatively high percentages of grass, sedge, *Artemisia* and *Thalictrum*. The constancy of *Artemisia, Helianthemum* and grass throughout the Lateglacial provides evidence that the vegetation of this dry region was always somewhat open.

The Region of the Alps (Fig. 2)

A substantial group of studies is available in southern Germany, mainly from young morainic landscapes. Beug's (1976) study of Sims-See is specially valuable because of the combination of detailed pollen and macrofossil analysis. At Sims-See (470 m) there are at first nearly 2 m of sediment with up to 80% herbaceous pollen. Pollen of grasses, sedges, *Artemisia, Helianthemum* and *Thalictrum* predominate. Leaves of *Dryas octopetala* and fruits and leaves of *Betula nana* occur. The herbaceous vegetation was invaded by a shrub assemblage of juniper (to over 30%), willow, and *Hippophaë*, all of which expanded their populations at about the same time. Slightly later *Betula* (tree-birch species, identified macroscopically) expanded to nearly 40% of the pollen, followed by *Pinus sylvestris*, identified by fossil needles, which ultimately became dominant with over 70% of the pollen. The diagram can be interpreted as an initial rather long period of stable alpine herbaceous vegetation, invaded successively by shrubs, tree-birch, and pine. Pine forest with birch was present until the Flandrian hazel rise. Evidence for climatic deterioration in the diagram is very slight. It appears unlikely that either a Zone Ic or III would have been identified on the internal evidence if this site had been taken in isolation. It is notable that the sediments assigned to Zone III, the period from about 11,000 to 10,000 BP, show evidence of slope erosion in their increased content of clay. The lithostratigraphic evidence for climatic deterioration is more striking than the evidence from the pollen diagram. Unfortunately, dates are not available.

Pechschnait (*c.* 600 m, Schmeidl, 1971) and Schwarzer See (1,896 m, Rausch, 1975) have similar records, but Kirchseeoner Moor (550 m, Rausch, 1975) shows a much more

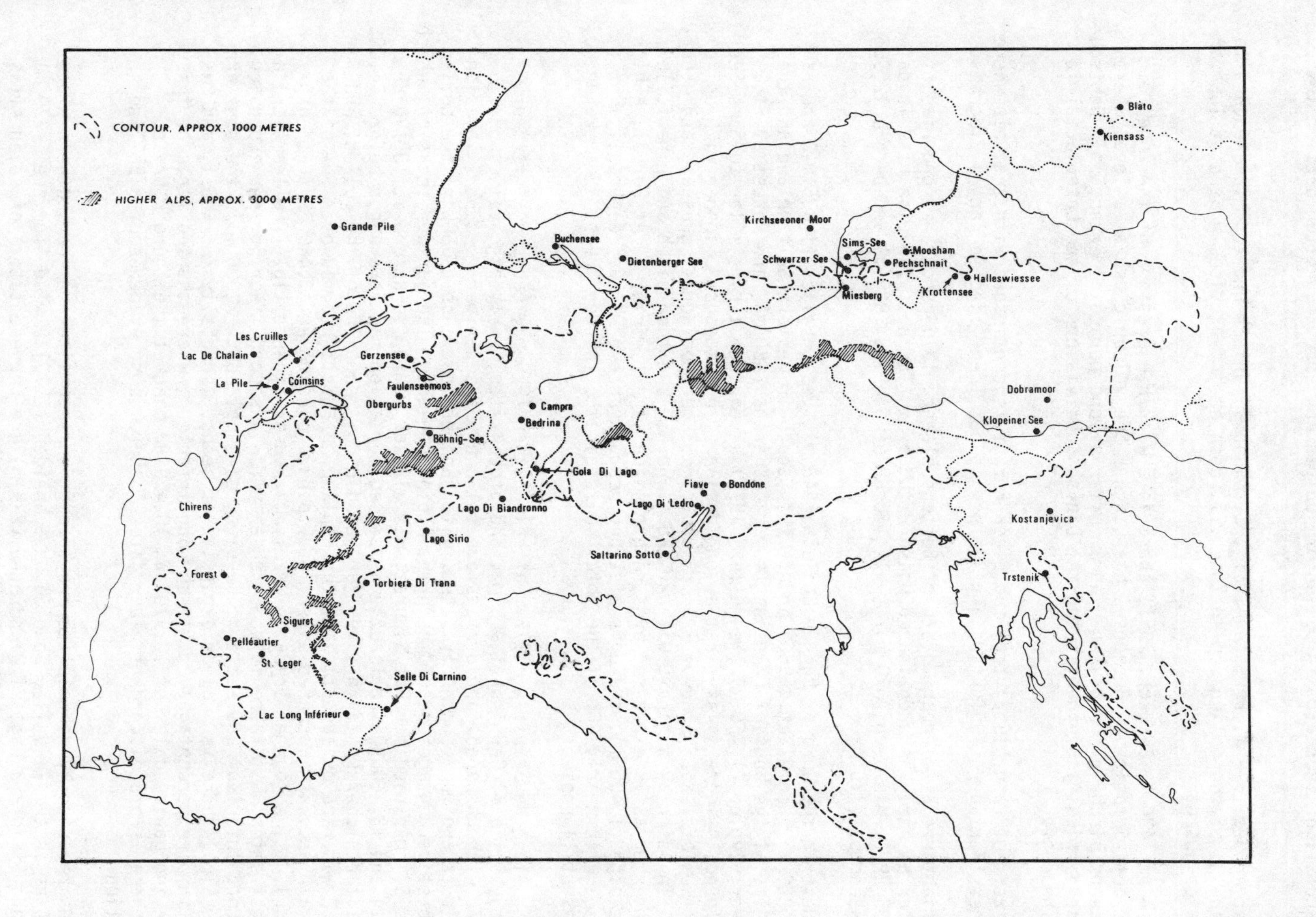

Fig. 2 Lateglacial sites in the Alpine region

clearly delimited Zone III, with a sedimentary change from algal gyttja to clay gyttja and slight increases in pollen of juniper and *Artemisia* after earlier low values.

Dietenberger See (679 m, Bertsch, 1961), also shows juniper and *Artemisia* each with more than 10% of the pollen sum in Zone III, both having previously been at low percentages or absent.

At Buchensee (430 m, Bertsch, 1961) there is a weaker expression of the Zone III climatic deterioration. Buchensee has Laacher See volcanic ash at a level where pine exceeds birch in the pollen diagram after a long period of expansion. Bertsch notes the difficulty of identifying the Danish subdivisions of the Lateglacial in the Alpine region.

Earlier work of Lang (1952) from the Black Forest is consistent with that elsewhere in southern Germany. Unfortunately juniper is not identified in Lang's diagrams.

In Austria three sites near Kufstein (Bortenschlager, 1976) show a strong rise of pine after an early herb-dominated phase with *Larix* (larch). The pine rise is dated to 13,035 BP at Miesberg (670 m), to 13,130 BP at Giering (820 m) and before 12,565 BP at Hasenmoos (770 m).

To the west in the Jura at Lac de Chalain (448 m), Les Cruilles (1,035 m), and La Pile (1,220 m) (Wegmüller, 1966) and farther south in western Switzerland and the French Alps at the bogs of Coinsins (480 m), Chirens (460 m), and Forest at St. Étienne-en-Dévoluy (1,460 m) (Wegmüller, 1977) a sequence very similar to Sims-See can be found. There is at first a long period of herbaceous dominance with up to 80% of *Artemisia*, *Helianthemum*, *Plantago alpina*, Chenopodiaceae, grasses, and other herbs. Invasion by juniper, willow, and *Hippophaë* and then birch and pine follows. The Younger Dryas is expressed by a re-expansion of *Artemisia*. At Forest, a high altitude site (1,460 m), there is a strong reappearance of herbs, especially *Artemisia*, with nearly 40% of the total pollen in the Younger Dryas. However, *Pinus* was still present.

In the SW Alps (de Beaulieu, 1977), at Pelléautier (975 m), Lac Long Inférieur (2,090 m), Selle di Carnino (1,905 m), and other sites, early high values of *Artemisia* with chenopods and grasses are followed by a juniper-*Hippophaë* invasion at the lower and more northerly sites, with juniper in excess of 50% of the pollen, the highest percentages recorded in the Alpine region. The juniper invasion, which occupied 50 cm of sediment at Siguret, was already coming to an end at 13,190 BP (Siguret) and 13,210 BP (Pelléautier), so sedimentation may have begun before 14,000 BP. At the higher southern sites juniper did not reach 10% of the pollen, and *Hippophaë* was rare or absent. In their place *Artemisia* reached over 50% of the pollen rain and was succeeded directly by pine expansion at 13,460 BP (Lac Long Inférieur). The Younger Dryas is weakly marked or imperceptible (Pelléautier) at the lowest sites, but there is some tendency for a second minor juniper expansion. At Lac Long Inférieur and Selle di Carnino, sites at about 2,000 m in the southern Maritime Alps, there is a clear expansion of herbs, especially *Artemisia*, and reduction in pine percentages. At Lac Long Inférieur, where the sediments of this time are inorganic, dates of 10,970 and 10,430 BP bracket the Younger Dryas, though there is one discordant older date. There is no evidence of a Bölling oscillation.

The sites reported by de Beaulieu are transitional to sites on the southern slopes of the Alps in the Po Valley studied by Schneider (1978). The 'Torfsee' (270 m) NE of Lago Sirio and Lago di Biandronno (239 m) are representative of seven sites studied. The profiles begin with high herb values, to 80%, with much *Artemisia* and grass. Juniper and *Hippophaë* are present throughout the phase at Lago di Biandronno, and their presence is well attested by several dates to before 13,290 BP.

Expansion of tree-birch, *Pinus cembra* and *Larix* took place after 12,630 BP followed by other pines (*cf. Pinus sylvestris*) at 12,000 BP. Traces of deciduous oak forest trees were present already at this time, and there is strong expansion in the Flandrian. At neither site is there evidence for climatic reversal. At one further site, however, Torbiera di Trana (360 m), there is clear reversion to high *Artemisia, Juniperus* and *Ephedra* and a fall-back in deciduous trees and pines to mark the Younger Dryas.

In the Lake Garda region, Grüger (1968) has shown at Fiave (654 m) and Bondone (1,550 m) that a sequence exists similar to that at sites farther west in the Po Valley, although there are differences in local detail. Tree-birch is present early at Fiave after a pioneer phase with the usual range of herbs and shrubs. Much of the Lateglacial is pine-dominated. Macrofossils of *Pinus sylvestris* occur at Fiave, and of *Larix* at both sites. At Bondone *Pinus* values lie at about 60% through much of the time before 11,530 BP. In a recession, pine falls to 50%, an effect achieved by an expansion in Cyperaceae. At a third site, Saltarino Sotto (194 m), there is a suggestion of climatic reversal in a brief juniper peak after birch and pine first expanded. Earlier work of Beug (1964) at Lago di Ledro (655 m) shows development generally similar to that at Fiave. At Fiave deciduous trees were already present in the Lateglacial and decreased in the 'Younger Dryas'. The same event, differently zoned, can be seen at Lago di Ledro. The zonation of Grüger, which places the Flandrian boundary at a decisive expansion of deciduous trees, is accepted here.

In the inner Alpine region, Eicher and Siegenthaler (1976) have shown that at Gerzensee (603 m) and Faulenseemoos (660 m) the vegetational sequence is similar to that in southern Germany and the Dauphiné. An early herb phase with abundant *Artemisia, Helianthemum* and chenopods is followed by pioneer shrubs (juniper-willow-*Hippophaë*), then tree-birch, then pine. Laacher See volcanic ash is found at both sites. The Younger Dryas is very weakly represented. There are small increases in *Artemisia* pollen, from 2% in the volcanic ash layer to 5% in the Younger Dryas at Faulenseemoos. $^{18}O/^{16}O$ values from the two profiles show that sharp climatic boundaries can be detected at the pioneer shrub phase and at the upper and lower limits of the Younger Dryas. The highest temperatures are recorded at the beginning of the pioneer shrub phase, with a slight decline through the forested period. At nearby Obergurbs (1910 m, Küttel, 1974) the record is similar. Very high pine percentages (80 - 90%) are maintained through the Younger Dryas, which is yet clearly marked by small percentage increases in *Artemisia,* chenopods, and other herbs.

In the eastern Alps in Carinthia, Fritz (1972) also records persistent very high pine percentages through the Younger Dryas, which is identified by a small rise in *Artemisia* and inflow of inorganic material.

In the southern Alps Müller (1972) has shown at Campra in the valley of Lucomagno that pine immigrated after a pioneer juniper-*Hippophaë* phase, and just before 12,890 BP retreated markedly in a subsequent regression in which herbs, especially *Artemisia,* returned to form more than 50% of the pollen. A date of 9,320 BP is associated with a birch rise, which initiated forest invasion once more. A further date of 9,100 BP, 330 cm higher in the profile, casts doubt on the younger date. Müller believes that the major climatic regression should be placed in the 'Piottino' cold phase recognised by Zoller (1960) and Zoller and Kleiber (1971 a, b) at Gola di Lago and other sites. The argument appears to depend largely on radiocarbon dates. Müller argues that a weakly expressed Younger Dryas regression is succeeded by a more severe 'Piottino' cold phase shortly after 10,000 BP at the transition to the Flandrian, but the dating evidence at Campra is hardly sufficient to sustain this opinion. It would be necessary to bracket both the 'Piottino' and 'Younger Dryas' sections of the profile with satisfactory dates to establish the case.

In a detailed diagram from Bedrina (1,235 m) Küttel (1977) has shown a single cold episode, consistent with the Younger Dryas but lasting until well after 10,000 BP. As Bedrina is the type-locality for the 'Piottino' cold phase, that name can no longer be used appropriately (Behre, 1978), and it is difficult to sustain the view that there were two cold phases between 11,000 and 9,500 BP.

Evidence for a prolonged Younger Dryas is reviewed by Behre (1978), who also considers evidence for a minor climatic warming in northern Germany (Friesland oscillation) and revertence during the earliest Flandrian. While the evidence for oscillations must be considered very carefully, excessive weight is often given to insufficiently numerous and well-established C14 dates and to relatively small fluctuations in pollen curves that might appear in a different light if pollen influx counts and macrofossil records were available. While Behre's case for a climatically complex record may prove correct in the light of further investigation, caution suggests that it would be wise to regard the question as not yet proven and to resist the proliferation of local 'oscillations' until there is very secure evidence.

Ireland and Comparison with Selected Sites in Western Britain (Fig. 3)

Watts (1977) and Craig (1978) have published new data on Ireland's Lateglacial vegetation and climate and have reviewed relevant literature. A very comprehensive review by Pennington (1977a) of sites in western Britain and detailed work on the Isle of Skye in western Scotland (Birks, 1973) and in the Scottish Highlands (Lowe and Walker, 1977) make it unnecessary to review the British literature fully. The objective here is to relate work in Ireland to the major conclusions emerging from British studies and to work in continental Europe. Reference may also be made to the work of Coope (1977) which discusses the significance of fossil beetles for interpretation of the Lateglacial climate.

In Ireland, sticky silts and clays, obscurely banded and often with layers of small stones, were deposited as the ice-sheet melted. Such deposits, which may be several metres thick, are normally sterile or have pollen at uncountably low densities. It is assumed that sedimentation took place in an unvegetated or very sparsely vegetated environment with great slope instability and active wastage of ice. There may have been a substantial time-lag between deglaciation and establishment of a vegetation cover. The earliest vegetation at most sites is characterised by *Rumex* (dock), *Salix cf. herbacea* (dwarf willow), grasses, and sedges. Sediments were still very minerogene. At Woodgrange (Singh, 1970), a flora with grass and much secondary pollen before the *Rumex* rise probably represents the earliest vegetation cover. The *Rumex-Salix* assemblage was displaced by invasions of *Empetrum* and often very abundant juniper, which have slightly different histories at different sites. *Empetrum* is abundant throughout the Lateglacial in western Ireland and western Scotland, where a type of arctic heath must have been present. The *Empetrum*-juniper assemblage is associated with high pollen influx values at Coolteen (Craig, 1978) and generally with organic or carbonate-rich sediments, which point to high lake productivity. There is sufficient evidence to suggest that the assemblage can be regarded as the period with highest July average temperatures in the Lateglacial. It is dated between 12,400 and 12,000 BP. It is difficult to determine the climatic succession that culminated in the juniper peak. One possible view is that a long slow invasion process took place, with progressive soil development and increasing complexity at the plant community level in a climate that was constant since the initial warming that caused the ice to melt.

Alternatively, the relatively sudden expansion of juniper at many sites, associated with clear indications of greater productivity in lakes, especially clearly seen at Coolteen (Craig, 1978), suggests that the climate may have become significantly warmer by about 12,400 BP. After 12,000 BP there is evidence for climatic deterioration. The juniper population collapsed, and silt layers are present, characteristically with some *Hippophaë*. From 11,800 to 11,000 BP the Irish landscape was

Fig. 3 Lateglacial sites in Ireland and Britain

grass-dominated. Although tree-birch was present, it played only a minor role in the vegetation. Sediments of this time are less organic than at the time of the juniper peak. The environmental implications are discussed by Watts (1977) and Craig (1978), but the climatic determinants of what must have been a prairie-like landscape with diverse herbs are difficult to determine. A very severe climatic deterioration followed after 11,000 BP. A peak of Cruciferae pollen indicates the first stage of soil erosion, but it is followed by vegetation in which *Artemisia* is characteristically associated with diverse Caryophyllaceae, saxifrages, *Armeria, Sedum rosea, Koenigia*, and other herbs. The sediments are poor in pollen, often contain abundant *Salix herbacea* leaves, and are very inorganic. Stones are frequent. There is clear evidence for extensive snow-patch vegetation and solifluction. The Flandrian then begins as the grass curve again rises, followed by peaks of juniper and birch. It should be noted that the beginning and end of the *Artemisia* phase, which is the equivalent of the Younger Dryas, appear to be rather abrupt. Although the grass phase (Alleröd equivalent) seems to represent a climatic decline after the early juniper peak, the *Artemisia* phase represents a further serious decline that may have taken place over a very short period.

In Britain, Pennington (1975a, 1977a) has published relative and absolute diagrams from Blelham Bog and Low Wray Bay of Windermere in the Lake District. They are similar to the record from Ireland in some respects but differ importantly in detail and in the radiocarbon ages attributed to events. The early *Rumex* zone is dated to before 13,000 BP at both sites, with basal dates in excess of 14,000 BP. A juniper peak lies between 13,000 and 12,500 BP and is followed by abundant birch between 12,500 and 12,000 BP. The period between 13,000 and 12,000 BP is assigned to a 'Bölling chronozone'. A climatic deterioration between 12,000 and 11,800 BP is identified on the basis of a steep fall in pollen concentration. Between 11,800 and 11,000 BP birch and juniper are again abundant, and after 11,000 BP the familiar 'Younger Dryas' *Artemisia* flora appears. The British record differs from the Irish in the much larger role of tree-birch, both between 12,500 and 12,000 BP and again between 11,800 and 11,000 BP, where there is also a juniper recovery in the equivalent of the Irish grass-dominated flora. The Lake District appears to have had birch woodland or forest when Ireland still had open landscapes. The dates are also consistent with the Irish record, but the extension of a Bölling chronozone back to 13,000 BP and the early date for the first vegetation are much older than anything recorded from Ireland and seem old even in comparison with central Europe. As the number of dates from Ireland is insufficient, further investigation may show that the juniper peak and earlier floras are older than has been assumed. However, the 14,300 and 14,550 BP dates recorded from Blelham Bog and Low Wray Bay, together with the basal date of 14,468 ± 300 BP from a kettle in North Wales (Coope and Brophy, 1972), appear to be the oldest recorded from any glaciated region and may be spuriously old because of undetected sources of old carbon. Confirmatory evidence from another region is desirable. Scottish sites such as Cam Loch (Pennington, 1975a) resemble the NW Irish site of Glenveagh (Watts, 1977) in their predominantly herbaceous vegetation with high percentages of *Empetrum*, indicating the presence of heaths throughout the Lateglacial. There is a very great wealth of regional detail in sites from Britain; not surprisingly western and SW sites (Seddon 1962; Dickson *et al*., 1970; Brown, 1977) most resemble the Irish record. The Lake District and sites in eastern Britain have a much greater representation of birch and seem to have been wooded through much of the Lateglacial.

Europe East of the Alps (Figs. 4 and 2)

At Witów in central Poland (Fig. 4), where dune formation took place in colder phases of the Lateglacial, Wasylikowa (1964a,b) claims to demonstrate the presence of Bölling and Alleröd oscillations. Sediments referred to Bölling were being deposited about 12,300 BP. The profile can be explained by an invasion sequence of herbs with willows and dwarf birch (macroscopic), then *Hippophaë*, followed by tree-birch and pine. A Younger Dryas is marked by a fall-back in pine from high (to

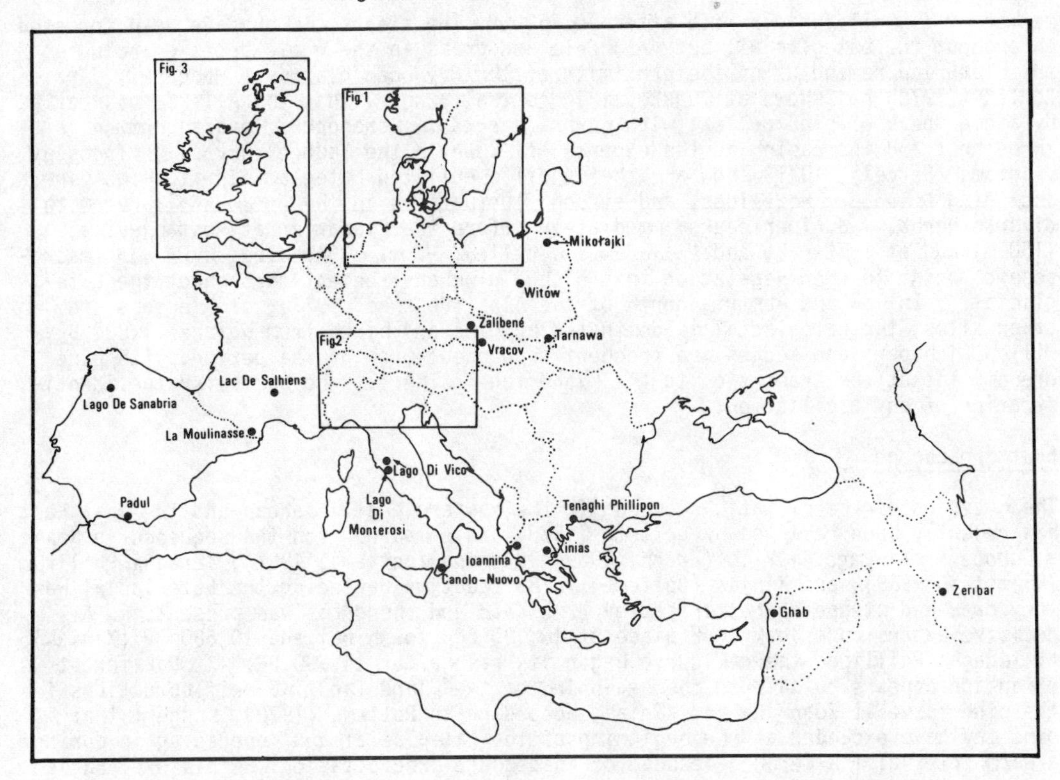

Fig. 4 Lateglacial sites in Southeastern Europe and the Near East

60%) values and relatively small increases in *Artemisia*, *Juniperus*, and chenopods. The evidence for a Bölling oscillation seems to depend on chronological equivalence more than unambiguous evidence in the pollen diagram for a climatic revertence after an early warm period. The relative movements of the pine and birch curves do not seem sufficiently sharply defined to serve as the basis for definition of an oscillation. The evidence for Younger Dryas at Witów is clear, although the percentage changes in the pollen diagram are slight. The same is true of Tarnawa (Ralska-Jasiewiczowa, 1972). Very small increases in *Artemisia*, juniper and chenopods dated between 10,880 and 10,390 BP provide satisfactory evidence for a clear but weakly expressed Younger Dryas. The recognition of a Bölling episode is not claimed.

At Mikolajki in northern Poland (Ralska-Jasiewiczowa, 1966) the older Lateglacial begins with fluctuating pine and birch values and high herbs, including *Artemisia* and *Helianthemum*. Dwarf birch, juniper, willow, and *Dryas* leaves were present macroscopically. At the end of this phase *Hippophaë* (leaf-hairs and pollen) and then tree-birch invaded. Pine (macroscopic evidence) expanded in the Younger Dryas (or late Alleröd) at the expense of birch. Clay-gyttja replaced limnic sapropel at this time, indicating upland erosion. A brief period of pine-decline with increase in herbs at the end of the phase of clay-gyttja sedimentation marks the end of the Younger Dryas and transition to the Flandrian. There is no evidence for a Bölling oscillation.

At Vracov in Czechoslovakia (Rybníčková and Rybníček, 1972) sedimentation in a bog profile began at about 12,000 BP. From early on, pine and birch predominated, with some 80% of the pollen. It is difficult to identify a Younger Dryas, but it may be marked by a small juniper rise after 10,760 BP. The area seems to have been forested throughout the Lateglacial, but *Artemisia* and grass in the whole profile including the Flandrian remind us of the proximity of the dry open plains of Hungary. Járai-Komlódi (1970) has shown at Dunakeszi in central Hungary that the Alleröd was marked by a high peak of pine pollen, with herbs (*Artemisia*, chenopods, grass) common throughout and increasing at the expense of *Pinus* in the Younger Dryas. Yugoslav diagrams (Sercelj, 1971) show a rather undifferentiated Lateglacial with high pine, including *Pinus cembra* values, and strong fluctuations in the *Artemisia* curve with diverse herbs. Juniper peaks immediately before the Flandrian at Kostanjevica (150 m) and at Trstenik, and *Pinus cembra* pollen, *P. mugo* needles, and *Dryas* (macroscopic) point to open vegetation in a high alpine environment throughout the Lateglacial. In lowland Austria north of the Alps (Peschke, 1977), at Kiensass and other sites, the Lateglacial is dominated by pine and birch from before 11,800 BP. Willow, juniper, and sedges are frequent throughout much of the period, with pine predominant at the transition to the Flandrian. There is no basis for the identification of any oscillation.

Southern Europe (Fig. 4)

The question of the climatic sequence in the eastern Mediterranean and the Near East has recently been reviewed by Bottema (1978) and compared with the sequence in N and W Europe. At Lago di Vico (Frank, 1969) Ioannina (Bottema, 1967), Tenaghi Phillipon (Wijmstra, 1969), and Xinias (Bottema, 1978) the time preceding the Lateglacial was very dry, and steppe characterised by *Artemisia* and chenopods was present. A decisive expansion of oak took place at 10,190 BP (Ioannina) and 10,680 BP (Xinias). At Tenaghi Phillipon the oak curve began its rise after 14,600 BP. Oak forest expansion appears to date to the beginning of the Flandrian. Some fluctuations in the pine curve at Ioannina and Xinias, according to Bottema (1978), suggest that pine may have expanded at the beginning of the Lateglacial and receded again during a warm interval ('Alleröd') because of inadequate precipitation. This explanation does not appear satisfactory, because at Tenaghi Phillipon a similar pine recession with high chenopod percentages is dated from before 14,600 BP and after 16,360 BP, dates much too old to be related to either the Alleröd or Bölling of N Europe.

At Lago di Monterosi (Bonatti, 1966, 1970), there is also evidence for 'Full-glacial' (Late Weichselian) *Artemisia*-steppe, and this has recently been confirmed by Grüger (1977) at Canolo Nuovo in Calabria, where herbaceous vegetation with very little tree pollen was still present at 14,000 BP. At Lac de Creno (1,280 m) in Corsica, Reille (1975) has shown that an *Artemisia* and grass-dominated herb vegetation was present until 10,420 BP. Claims that an Alleröd horizon is recognisable at the base by an increase in tree pollen may perhaps be explained by rebedded pollen.

In the Ghab Valley of western Syria (Niklewski and van Zeist, 1970) oak invaded herbaceous communities before 10,080 BP, while at Zeribar in western Iran (van Zeist, 1967) a decline of *Artemisia* at 13,650 BP resulted in an expansion of the chenopod curve, with a delayed and slow expansion of oak until the mid-Flandrian. This sequence may indicate increased temperature after 13,650 BP but only a very slow increase in precipitation.

In S Spain at Padul (Florschütz *et al.*, 1971) oak expanded before 13,000 BP into a vegetation earlier dominated by pine and herbs with some oak. A secondary expansion of pine took place before and around 10,470 BP, with oak predominant in the Flandrian. There is a clear suggestion of Lateglacial climatic oscillation, but the oak rise may have begun before 15,000 BP, judged by the spacing of the available radiocarbon dates, and the sampling interval is too wide in the pollen diagram to yield sufficient detail for a satisfactory interpretation. At Laguna de las Sanguijuelas, a marshy area beside Lago do Sanabria (also called Lago de Villachica) in NW Spain, Menéndez Amor and Florschütz (1961) show a major increase in birch and a decrease in pine, *Artemisia*, and other herbs at 11,585 BP. Before this, pine and herbs predominated except for a birch peak at 13,700 BP. The birch peak at 11,585 BP is used to identify the beginning of the Alleröd. Successive peaks of birch, pine, and trees of deciduous forest are referred to the Alleröd, and a fall in pine and oak forest with a new birch expansion to Younger Dryas. The available dates make the interpretation possible, but as at Padul the level of information is simply insufficient to permit detailed interpretation. The Spanish sites, for the moment, are rich in potentially interesting information, which has not yet been realised adequately.

Jalut (1973) has published a pollen diagram from La Moulinasse (1,330 m) in the eastern Pyrenees that shows a pine invasion into herbaceous vegetation considerably after 13,600 BP. There is a credible suggestion of a Younger Dryas reversion, but detail is lacking. This site and Lac des Salhiens (1,220 m, Lang and Trautmann, 1961) may be compared more appropriately with the western Alps than with the Mediterranean. Lac des Salhiens unfortunately lacks *Juniperus* determinations and radiocarbon dates. It has a very long and complex history of herb-dominated communities before birch and oak expansion indicate the beginning of the Flandrian. Further study is very desirable at this interesting site.

SUMMARY

Regional Variation

In the Alpine region there is no single site that provides clear evidence for two Lateglacial climatic oscillations. Many authors have attempted to find a Bölling equivalent but have either acknowledged failure or made unconvincing efforts to subdivide the older parts of their diagrams. Detailed diagrams from the Alps usually show long pioneering successions. Herbaceous vegetation, usually rich in *Artemisia*, *Helianthemum*, and chenopods is invaded by juniper, willow and *Hippophaë*, often more or less simultaneously, then tree-birch, then pine (*Pinus sylvestris*). At west Alpine sites the rise of pine to dominance over birch occurs shortly before the Laacher See volcanic ash horizon. Pine and birch remain present during the Younger Dryas, often with a decrease in birch. A new rise in birch marks the transition to the Flandrian. A Younger Dryas horizon may not be identifiable, but it may be marked by an increase in *Artemisia* and juniper pollen, and at some sites by inflows of inorganic sediment. It is much more marked at some sites at higher altitudes. The Alps show great diversity regionally. Juniper is an important plant in the western Alps and is present in herbaceous vegetation from the first records in the Po Valley. North of the Alps it is a pioneer after a period of herbaceous vegetation, and farther east and into the lowlands of eastern Europe it is much less significant in the vegetation. In the eastern Alps pine invades herbaceous vegetation directly without a pioneering shrub or tree-birch stage, while in the Po Valley *Pinus cembra* and *Larix* were early invaders. It can be concluded that the Alps show clear evidence for a 'Younger Dryas' climatic deterioration at many but not all sites. The forest did not disappear but seems to have opened up, especially in southern sites and at high altitudes, to allow vegetation with juniper and *Artemisia* to develop locally. There is evidence for slope erosion in the form of increased supplies of minerogenic material to lake basins.

In the North European Plain and Southern Scandinavia, the evidence for a 'Younger Dryas' climatic reversal is clear. It takes the form of an opening of the birch and pine forests to allow *Empetrum* heath to develop in sandy regions of North Germany and the Netherlands. There is evidence that coversands were moved at this time. In Denmark birch seems to have persisted in the Younger Dryas but declined in quantity. Herbs increased and inorganic materials were deposited in lake basins. At some Swedish sites (Åkerhultagöl) the decline in tree pollen from about 11,000 to 10,200 BP is very marked. However, the floristic changes are more quantitative than qualitative.

The problem of the Bölling oscillation remains. It is not satisfactorily demonstrated at Bölling because of the absence of dates and the dependence on change in the birch and sedge curves only. It would be necessary to re-study Bölling at Iversen's original locality and date it there if it is to be acceptable as a stratotype or to name a chronozone. Evidence of a Bölling-type oscillation exists at Usselo and at Seck-Bruch (Dietz *et al.*, 1958), but the most convincing evidence comes from Gaterslebener See. Against that, there is a substantial number of sites in the same region that show little credible evidence for a Bölling oscillation (Glüsing, Westrhauderfehn, *etc.*). Mangerud (1970) has provided evidence for a minor ice readvance in western Norway shortly before 12,000 BP at Blomöy, but this is not necessarily associated with vegetation changes at a distance from the ice front. New studies of the Bölling type-locality are very desirable. It would also be of great interest to see a pollen influx study of the oscillation at Gaterslebener See to be certain that what appears to be an oscillation is not an artefact of low local pollen production.

In Ireland and Britain the 'Younger Dryas' episode is very strongly marked. There is clear evidence of renewed mountain glaciation in Ireland at Lough Nahanagan (Colhoun and Synge, 1979), at Windermere, where banded sediments between 11,000 and 10,500 BP give evidence for active ice in the mountain catchment (Pennington, 1977a), and in Scotland, where there was major glacier growth during the Loch Lomond Stadial (Sissons, 1974b, 1976a). The climate at this time must have been very severe. At many sites the sediments contain pollen at very low densities. The flora was greatly changed at many sites as compared with the preceding period, so that one can speak of a qualitative rather than a quantitative change. At many sites *Juniperus* disappears (*e.g.* Cam Loch, Ballybetagh). *Artemisia, Koenigia, Sedum rosea* and saxifrages are largely confined to this zone. At many sites sediments are extremely minerogenic and contain coarse erosional debris, including small stones. Ireland and Britain show much the strongest expression of the Younger Dryas deterioration in Europe.

The question of a 'Bölling' oscillation in Ireland and Britain is of considerable interest. Craig (1978) has shown at Coolteen that strong upland erosion took place after a period of stability and productivity in the landscape corresponding to an early juniper peak and dated shortly before 12,000 BP. At many Irish sites a band of inorganic material just after the juniper peak suggests erosion after 12,000 BP. Perhaps more striking is the general evidence of the flora that juniper gave way to grassland. It is possible to argue that the erosional phase represents a beginning of climatic deterioration, which continued through the 'Alleröd' from 11,800 to 11,000 BP, with an unstable climate suggested by strong fluctuations in birch, juniper, and herb curves at many sites. A 'Bölling' oscillation is claimed to have been identified at several sites in northern Britain by Walker and Godwin (1954), Oldfield (1960), Bartley (1962) and Pennington (1975a and b). Recently double maxima for woody plant pollen in the Lateglacial have been reported from two sites in Scotland, and the authors cautiously suggest the possibility of an equivalent to the Bölling oscillation at these sites (Walker, 1977; Caseldine, this volume). It is somewhat ironic that the best evidence for a 'Bölling' oscillation followed by a deterioration ('Ic') comes from Ireland and Britain, while the evidence from Bölling itself gives cause for doubt.

Timing and the Lateglacial Climatic Curve

Ruddiman *et al.* (1977) calculate that deglacial warming of the North Atlantic began at about 13,500 BP and that a reversal to cold conditions (the Younger Dryas) centred on 10,200 BP. There are no further records of climatic change in the ocean cores. It is difficult, however, to delimit and date events accurately because of vertical mixing of sediments by burrowing fauna. This results in blurring and smoothing of the climatic record and in reduction of the amplitudes of climatic oscillations with the greatest loss for the shortest oscillations. In principle, the land record should be more detailed and accurate.

A substantial group of dates from S Sweden (Berglund, 1976) shows that the oldest dates in the ice-marginal zone are about 13,500 BP. The dates were carried out on marine molluscs or marine vertebrate bone. The oldest date on a lacustrine sediment is 13,020 BP. The dates are consistent with deglacial warming earlier than 13,500 BP. Few dates of comparable age are known from Scandinavia, North Germany, or the Netherlands, but data from western Norway (Chanda, 1965; Mangerud, 1970), show deglacial warming by 13,000 BP with a 'Bölling' birch expansion at 12,650 BP, corresponding to a similar date in Berglund's Håkullsmosse site. Mangerud provides evidence for minor ice-margin readvances slightly before 12,070 BP and again at the time of the Younger Dryas.

In the Alpine region pine was expanding at about 13,000 BP at Miesberg and nearby sites (Bortenschlager, 1976). Pine was expanding at Lac Long Inférieur (de Beaulieu, 1977) at 13,460 BP, and a juniper expansion into herbaceous vegetation was ending at 13,190 BP at Siguret (de Beaulieu, 1977). South of the Alps at Biandronno (Grüger, 1968) trees were invading open vegetation after 12,630 BP. No dates older than 13,460 BP exist for tree expansion, although relatively long periods of herb dominance precede this time. There does not seem to be a satisfactory basis for identifying the end of glacial cold and the beginning of deglacial warming in herb profiles, and the tree expansions are presumably time-transgressive. A climatic warming to permit trees to expand as early as 14,000 BP or slightly earlier can be envisaged. Very early dates back to 14,600 BP for the beginning of re-vegetation in the English Lake District have been commented on.

There seems to be very general agreement that a Younger Dryas period dates to approximately 11,000 to 10,000 BP (termed the Loch Lomond Stadial in Britain - Gray and Lowe, 1977). Pennington (1977a) records that annual varves are present in Windermere from 11,000 to 10,500 BP, indicating active ice in high corries. At Lough Nahanagan in Ireland, Colhoun and Synge (1979) show that a glacier was present in the Younger Dryas equivalent, but that silts that formed over the glacier's moraines still contained an *Artemisia*-dominated flora, showing that the vegetation was unchanged for some time after the glacier melted out.

There also seems to be general agreement that a climatic deterioration took place between about 12,000 and 11,800 BP. It is rather clearly marked in Ireland and Britain, but the evidence in Denmark and Germany is somewhat inconclusive. There is no convincing episode from this time in the Alpine or South European record. It is not certain that the period from 11,800 to 11,000 BP had a homogeneous climate. In Ireland and Britain it appears to have had a colder climate than the period before 12,400 BP, but very strong fluctuations in pollen influx at Blelham Bog (Pennington, 1975a) suggest that the climate may have varied considerably. At Blelham there is a forked curve for deposition rates of birch in the Alleröd chronozone. Pennington comments on its resemblance to Iversen's double peak of birch in the Alleröd of Denmark. She also observes that the birch and juniper curves at Blelham Bog at this time "suggest an interplay between birch and juniper within a fluctuating environment" (p.163).

The course of climatic change is often represented graphically by curves to show changes in mean July temperatures. Obviously many more factors may be involved, such as length of growing season, winter temperature, frequency of extremes of temperature or precipitation, mean precipitation, windiness, *etc.*, so that July mean curves may be a very misleading over-simplification of a complex picture. Iversen (1954) stresses the over-riding importance of temperature, and it seems certain that warming and cooling climatic trends can be identified and a curve drafted to express this.

Little attention is usually given to the form of the curve, because it is assumed that climatic change is gradual since the plant response is expressed as population growth. If it is sudden, however, changes may take place over a few decades, and the proper form of the curve is a steep gradient by which one steady state is changed to another. Alternate possible forms of representation of a climatic curve for the Lateglacial are presented in Fig. 5.

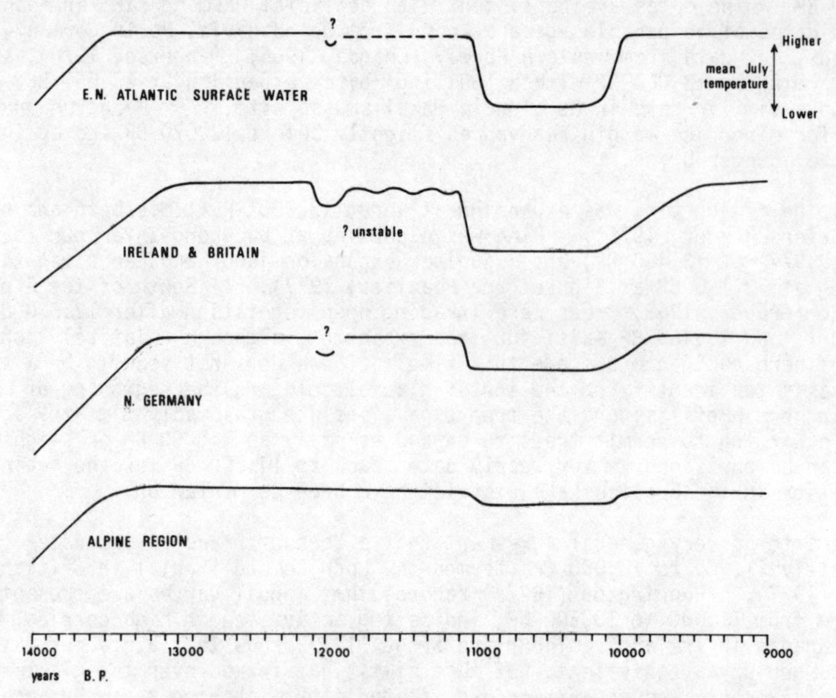

Fig. 5 Schematic Lateglacial climatic curves, inferred from records of ocean surface temperatures and changes in vegetation from three regions of Europe

It may be questioned whether the Lateglacial should be regarded as an interstadial. An interstadial may be defined as an episode during a period of glacial climate when sufficient climatic warming takes place for a temporary establishment of plant formations with higher climatic demands than indicated by the vegetation record in the bracketing sediments. The Lateglacial resembles an early interglacial warming in its long successional build-up of vegetation types of increasing complexity. The Younger Dryas equivalent represents a real climatic cooling and change in NW Europe, but it is merely a delaying factor in a little-changed succession in Central Europe. The 'Bölling' and 'Alleröd' of the Danish area and their equivalents in the British Isles were separated by a short cold phase in NW Europe that is not duplicated in the Alps or farther east and south. The two 'warm' phases are therefore distinct in some areas but united in others. The Lateglacial is therefore a complex and eventful climatic episode of a unique kind that may be classified with interstadials but that is distinct from them in many respects.

The Unique Character of the Lateglacial

The Flandrian appears to be the only interglacial cycle immediately preceded by a complex climatic event like the Lateglacial. There is no evidence for such an event before the Eemian or Holsteinian Interglacials known from a large number of profiles, nor, so far as the author is aware, from the less well known earlier interglacials. Interglacial cycles of vegetation normally reveal pioneering succession of herbs and shrubs to forest trees, but without any revertence. The Lateglacial has the character of an interglacial climatic/vegetational cycle that made a false start and had to begin again. It is powerfully expressed in the vegetation record of NW Europe and more weakly elsewhere in Europe. In eastern North America no revertence has been found, and the flora seems to have progressed steadily from pioneer stages to broadleaved forest without interruption (Wright, 1971). The Younger Dryas revertence is recorded in cores from the eastern but not the western North Atlantic. Evidence for Lateglacial climatic oscillations elsewhere that may be correlated in time with those in Europe is not satisfactory (see Wright, 1977, for critical review). Records of such oscillations may represent purely local events in high mountain regions that are quite independent of the European climatic episode. For example, Schreve-Brinkman (1978) has clearly shown a cold phase at El Abra (2,570 m) in Columbia, beginning at 11,210 BP.

From the palaeoclimatic point of view, the Lateglacial is of great interest. Here is a major climatic episode known from only one part of the world and recorded locally in the ocean record. One wonders whether there may be other local records of unrecognised similar events elsewhere in the world and further back in the Quaternary stratigraphic column. It is commonly believed that significant climatic events should find a reflection in the stratigraphic record throughout the world. We may have to distinguish between genuinely world events, such as the beginning of warming in the oceans at the end of the last major glaciation, *e.g.* termination 1 of Emiliani (1966), and major events of lower rank, such as the Younger Dryas cold episode, strongly expressed in a large continental area and the adjoining ocean, but unrecorded elsewhere.

Ice-front Variations of Different Parts of the Scandinavian Ice Sheet, 13,000-10,000 Years BP

Jan Mangerud

(University of Bergen)

ABSTRACT

Ice-front variations of the Scandinavian Ice Sheet from 13,000 to 10,000 years BP are summarised with the main emphasis on the Allerød and Younger Dryas chronozones. In Finland, Sweden and eastern Norway there was a great net retreat, in SW Norway the ice-front oscillated back and forth, and in the NW part of southern Norway the ice-front retreated during the Allerød and halted during the Younger Dryas. The differences are interpreted in terms of different topography and therefore different glacial responses to climatic changes.

In the present paper ice-front variations of the Scandinavian ice sheet from 13,000 to 10,000 BP are discussed by means of time-distance diagrams. First the basis and the accuracy of each curve are discussed. Thereafter the curves from different parts of the ice-front are compared. The positions of the profiles are indicated in Fig. 1, and all curves are shown in Fig. 2.

Estonia - Western Finland

The curve for this area is constructed from data given in Donner (1978). For the Luga-Haanja moraine an age of 13,200 - 13,000 BP was accepted, and for the Neva-Pandivere moraine 12,200 - 12,000 BP. The average distance between the two moraine systems is about 140 km.

The ice margin probably withdrew to the south coast of Finland at 11,800 or 11,600 BP. The latter date has been adopted for the relevant curve. However, Donner points out that this age implies that northern Estonia was deglaciated earlier than the commonly accepted age of 11,200. Thus the age suggested for the Neva-Pandivere moraine may also be too young.

On the basis of correlations of C14-dated pollen-zone boundaries, Donner concluded that the Salpausselkä III moraine was deposited immediately before 10,100 BP. The duration and relative age of each of the three Salpausselkä moraines is based on the varve chronology of Niemelä (1971). Possible errors for the Younger Dryas part of the curve are thought to be minor.

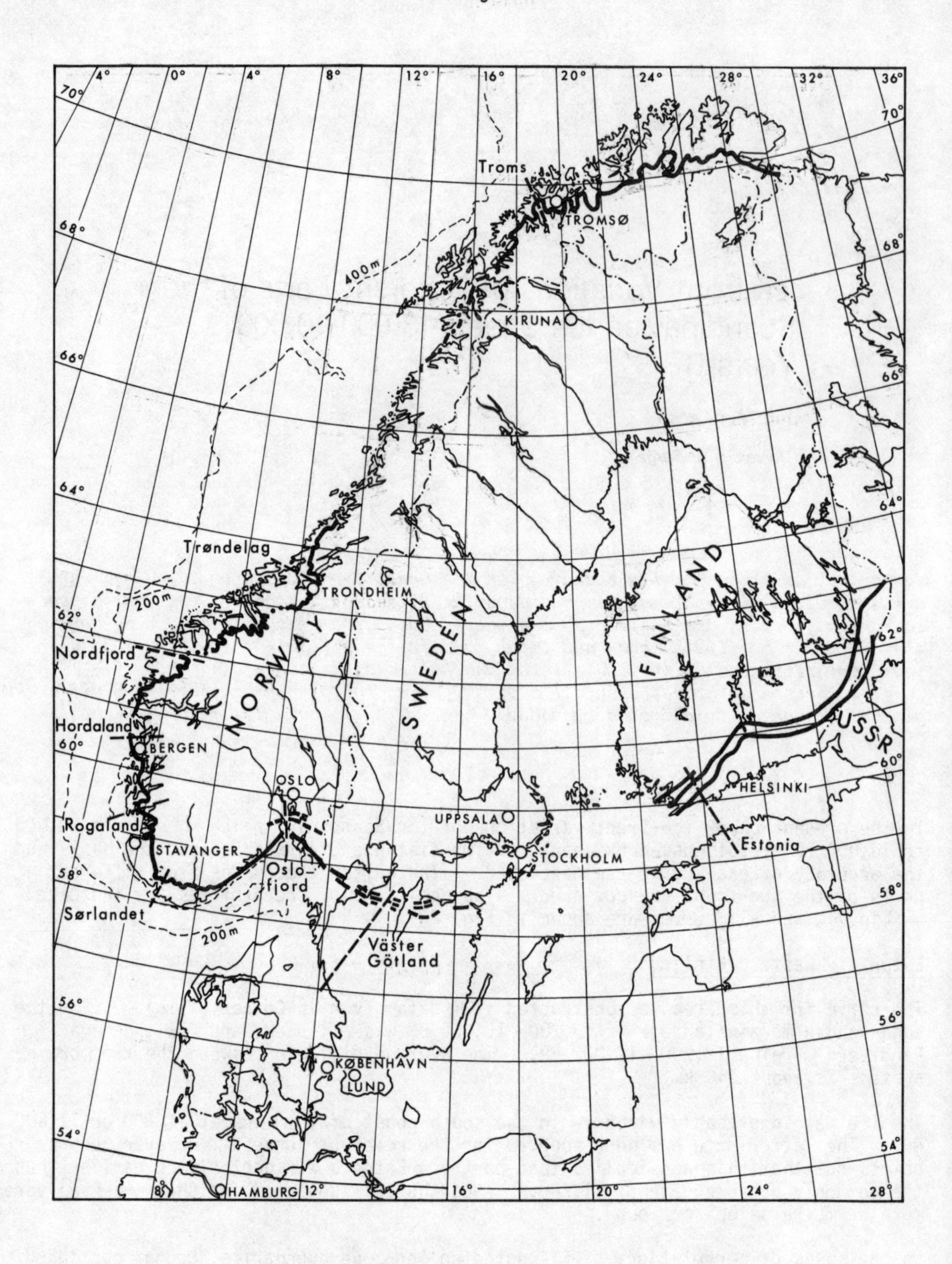

Fig. 1 End-moraines of Younger Dryas age around the Scandinavian Ice Sheet. The positions of the profiles given in Fig. 2 are indicated.

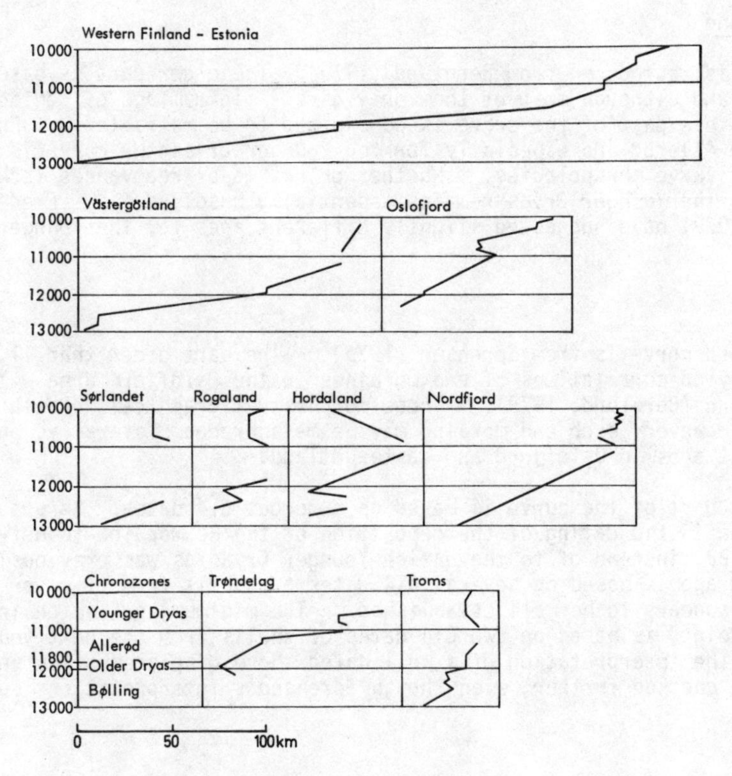

Fig. 2 Time-distance diagrams for the position of the inland ice front.

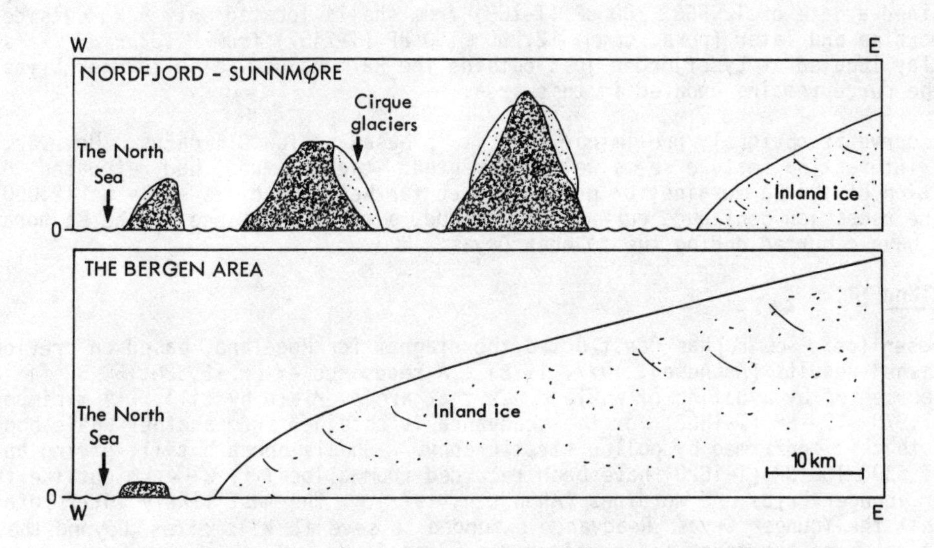

Fig. 3 A simple model showing the difference in glaciation types between Nordfjord-Sunnmøre and the Bergen area (Hordaland) during the Younger Dryas.

Västergötland

This curve is reproduced from Berglund (1979). The older part is based on numerous C14 dates, and although most of them only give a minimum age of deglaciation at each site, this part of the curve is considered to be reliable. For the younger part of the Alleröd and especially for the Younger Dryas the curve is mainly based on floating varve chronologies. Whether or not major readvances took place in this area during the Younger Dryas remains essentially unsolved. Fredén (1978) and Sørensen (1979) have suggested slightly different ages for the Younger Dryas end moraines.

Oslofjord

The Oslofjord curve is from Sørensen (1979). The part older than 11,000 BP is based mainly on correlations of end moraines in the Oslofjord area with moraines in Västergötland (Berglund, 1979). These morphostratigraphical correlations seem convincing. However, each end moraine may be metachronous laterally, and may thus be of different ages in Oslofjord and Västergötland.

The younger part of the curve is based on numerous C14 dates. Sørensen's major new contribution is the dating of the deposition of the Ra moraine to between 10,800 and 10,600 BP, instead of to the entire Younger Dryas as was previously assumed. The maximum age is based on several C14 dates of shells from till, or sub-till sediments, and appears to be well established. The minimum age, which in this case is a crucial point, is based on two C14 dates of shells from the next younger moraine (Mona) and the interpretation of a well dated shore displacement curve. This age needs to be checked further, even though Sørensen's interpretations seem likely.

Sørlandet

This curve is constructed from Andersen's data (1960, 1979). He assumes an age of 13,500 BP for the Lista moraine, on the basis of a C14 date of 13,150 ± 300 BP (T-149 B) from gyttja deposits on the proximal side of the moraine. Andersen (1960) obtained a date of 12,550 ± 200 BP (T-168) from shells located only 9 km outside the Ra moraine and later (pers. comm) 12,250 ± 150 BP (T-2467) from *Mytilus edulis* shells in clay located at Lysefjorden just outside the Ra moraine. The Younger Dryas part of the curve remains undated in this area.

This curve is obviously preliminary, as it is based on few C14 dates. However, one very interesting feature seems well documented: the ice-front had retreated to the position of the Ra moraine, or probably even further inland, as early as 12,000 BP. If the recession continued during the Alleröd, a major readvance to the Ra moraine must have occurred during the Younger Dryas.

Rogaland

Anundsen (pers. comm.) has constructed the diagram for Rogaland, based on previously published results (Anundsen, 1977, 1978). A readvance after 12,380±150 BP (T-1621)) is documented by a dating of whale bones that are overlain by till. A minimum age of 11,970 ± 100 BP (T-1883) for the readvance is obtained from another whale bone date that is confirmed by pollen stratigraphy. Shells beneath till, giving an age of 11,630 ± 100 BP (T-1620) have been recorded from a locality 3 - 4 km outside the major Younger Dryas end moraines (Anundsen, 1977). The most likely interpretation is that the Younger Dryas Readvance extended to several kilometres beyond the position where the major end moraines were deposited.

For the Younger Dryas, Anundsen concludes that the ice-front reached its outermost position prior to 10,720 ± 180 BP (T-995). This is based on a date of organic matter in clay which he assumes was deposited close to the ice-front.

Hordaland

This curve is from Mangerud (1977), and is based on approximately forty C14 dates. A readvance at about 12,000 - 12,400 BP (probably 12,200 - 12,300 BP) is documented by several C14 dates of shells, whale bones, and wood below till. The retreat inland which followed during the Alleröd is demonstrated by many dates of shells in till or below till. During the major Younger Dryas readvance the glacier overrode the Alleröd deposits, and therefore minor halts or readvances during the Alleröd and early Younger Dryas would have been obscured. It appears that the ice did not reach the outermost Younger Dryas position simultaneously along the entire end moraine (Aarseth and Mangerud, 1974). However, even along the younger part of the moraine, the readvance may have culminated closer to 10,500 BP than to 10,000 BP as previously concluded by Mangerud (1977), since the latter age is based on only two C14 dates.

Nordfjord

This curve is from Mangerud *et al.* (1979), and is extended 40 km offshore on the basis of a C14 date of 13,350 ± 340 BP (T-2708) of shells from glaciomarine clay that lies outside a submarine end moraine (Nordvik and Holtedahl, pers. comm.). However, the curve prior to 11,000 BP must be regarded as tentative, since it is based on only two C14 dates. The Younger Dryas part is also based on few dates, but data from key stratigraphical sites suggest this part to be more reliable. A readvance, which apparently did not deposit end moraines, is bracketed between 10,750 ± 140 BP (T-2304: a date based on shells in till) and 10,440 ± 170 BP (T-645). The huge Younger Dryas end moraines, named the Nor Moraines, were deposited after the latter date.

Trøndelag

This curve is based on dates from a large area and therefore there are problems with lateral correlations. The distances used in the diagram (Fig. 2) are partly measured south of the profile indicated on the map (Fig. 1).

From Kristiansund, Kraemer (in Gulliksen *et al.*, 1978) described shells in till yielding a date of 12,090 ± 100 BP (T-1805), thus indicating a younger readvance over this site.

Lasca (1969) dated an ice-front deposit to 11,300 BP by means of several C14 dates. He assumed the moraine to be a part of the Younger Dryas end moraine system, while Andersen (1979) has demonstrated it to be an older moraine.

Three C14 dates relate to a large ice-front delta at Heimdal, near Trondheim (Reite in Nydal *et al.*, 1972, p.528; Sollid and Sørbel, 1975). A whalebone was found in the lower part, or possibly below the delta deposits, and the C14 age is reported as 11,290 ± 190 BP (T-787). However, the δC13 was not measured and corrections for neither isotopic fractionation nor reservoir age were applied (Gulliksen, pers. comm.). Assuming a δC13 value of -16.5°/oo PDB, and a reservoir age of 440 years (Mangerud and Gulliksen, 1975), these corrections would imply a younger age by about 300 years, that is 10,990 ± 190 BP. Two dates from shells in a clay above the delta deposits are 10,230 ± 130 BP (T-786) and 10,150 ± 100 BP (T-754). Sollid and Sørbel (1975, 1979) have shown that the Heimdal delta belongs to the prominent Younger Dryas end moraine system, which is now mapped as nearly continuous along the coast to the Ra moraine in Oslofjord (Fig. 1).

Kjemperud (1978) obtained six C14 dates from three localities about 25 km proximal to the Tautra moraine, which is correlated with the Heimdal delta. This indicates that the sites were deglaciated at about 10,300 - 10,500 BP.

Troms

This curve was constructed by T.O. Vorren (pers. comm.), based on data from Andersen (1968), K.D. Vorren (1978), T.O. Vorren *et al.*, (1978) and T.O. Vorren and Elvsborg (1979). Two parts of the curve are well established: the Skarpness end moraines were formed at about 12,000 - 12,400 BP, and the younger part of the Tromsø-Lyngen end moraine complex were formed during the Younger Dryas, probably at about 10,200 - 10,600 BP. Few dates are available for the rest of the curve.

The curve is not extended to offshore areas. Nevertheless, an early deglaciation of the continental shelf, as concluded by T.O. Vorren *et al.*, (1978) is assumed. Rokoengen *et al.*, (1977) assumed a considerably later deglaciation of the continental shelf, mainly on the basis of the interpretation of fossil-bearing overconsolidated clays being basal tills. The dates from deposits on land strongly indicate that the model of Vorren *et al.*, is correct, but the interesting problem of the overconsolidation remains unsolved.

DISCUSSION AND CONCLUSIONS

Though the total period considered above is comparatively short (*c.*3,000 years), a significant proportion of the deglaciation of the last Scandinavian Ice Sheet occurred within this time. For exact chronostratigraphical correlations within such a short period, detailed stratigraphy and precise datings are necessary. Our present state of knowledge does not allow any detailed comparisons of the curves. However, some major trends and regional differences seem well documented.

1. From 13,000 to 10,000 years ago there was a major retreat of the eastern flank of the Scandinavian Ice Sheet, and also in some areas in the west, while the ice-front oscillated back and forth on the coast of SW Norway where there are high and extensive mountain plateaux along the coast.

2. A halt or a readvance at 12,000 BP, or slightly earlier, is recorded in most areas. This was probably caused by a climatic event, the differences in time and extent being the result of different response mechanisms and response times in individual parts of the ice sheet.

According to the previously used climatostratigraphical subdivision, these halts and readvances could all be regarded as Older Dryas events, given different amplitudes and time-lags for the response of different biological and physical processes to climatic changes (Mangerud, 1970). However, several of these events did not occur within the Older Dryas Chronozone, as defined by Mangerud *et al.*, (1974).

3. For the Alleröd and Younger Dryas chronozones more details are available, and three different types of glacial response can be distinguished:

A. In the eastern area (the curves for Finland, Västergötland and Oslofjord) there was a major net retreat from 12,000 to 10,000 BP. There was also a net retreat during the Younger Dryas, though the glacier front oscillated somewhat. Trøndelag is also included in this group, as the relevant curve is very similar to that from Oslofjord. Similarity in response is probably related to the influence of topography, as there are extensive lowlands in Trøndelag and in the eastern areas.

B. The curves for SW Norway (including Sørlandet, Rogaland and Hordaland) together with Troms in northern Norway show that the ice-fronts retained nearly the same positions at 12,000 BP and 10,000 BP. For parts of Sørlandet the ice-front may have been even further inland at 12,000 BP than at 10,000 BP.

The retreat during the Alleröd in Hordaland was of the same order of magnitude as retreat in Oslofjord and Trøndelag. However, a major readvance occurred during the Younger Dryas in Hordaland but not in Oslofjord nor in Trøndelag. It is also assumed that a similar major readvance occurred in Sørlandet and Rogaland, even though this is not proven. In Troms the glacier reactions may have been slightly different.

C. The response of the ice sheet in the NW part of southern Norway, represented by the curve from Nordfjord, was similar to that of Group A, except that no retreat took place during the Younger Dryas. Outside the ice-sheet, however, a great number of cirque glaciers were formed (Fig. 3). This was a common occurrence in the Troms area (Andersen, 1968).

The variation in glacier response discussed above could be a result of climatic differences, especially differences in precipitation. That is, however, not necessarily the cause, or at least not the only cause. An earlier explanation (Mangerud, 1970), that the differences in ice-front variations between Hordaland and Västergötland result from glaciological differences, is strengthened by the additional information presented in this paper.

Southeastern Scandinavia, with response type A, is a lowland area. There were relatively long distances between the accumulation areas and the ice-front, and there was divergent ice-flow throughout this region, such that any net accumulation became spread out over an extensive ice-front. The result was that a considerable increase of accumulation was necessary to produce a major readvance, and the response time was long.

In SW Norway (type B) there are extensive mountain plateaux close to the coast, only dissected by narrow valleys and fjords. During the Younger Dryas these plateaux were accumulation areas, and there were convergent ice-flows through the valleys and fjords, resulting in a rapid and large readvance. Even during the Younger Dryas maximum there were but short distances between the accumulation areas and the ice-front. Similarly ice retreat during mild phases (in this case the Alleröd) was rapid in this area due to calving in the deep fjords. Therefore, in this area, topography caused the ice-sheet to respond rapidly and with large ice-front oscillations to short-lived climatic fluctuations. This interpretation is strongly supported by the similarities between the curves from Trøndelag and Oslofjord, as Trøndelag has climatic similarities with SW Norway and topographic similarities with the Oslofjord area. The difference between SW Norway (type B) and the area immediately further north (type C) can hardly be explained by climatic differences. There is, however, an important topographical difference (Mangerud *et al.*, 1979). As described above there are large mountain plateaux in the southern area, while the topography in the northern area is much more dissected. In the latter area many local glaciers were formed, but they did not coalesce into one ice sheet (Fig. 3).

4. Even though there is more than one end moraine of Younger Dryas age in some areas, in southern Norway there is only one major end moraine of that age, which is mapped nearly continuously from Oslofjord to Trøndelag. However, this one end moraine, often called the Ra moraine far outside the Oslofjord area where the name originated, is demonstrably metachronous. The Ra moraine in the Oslofjord was formed before 10,600 BP and the Herdla moraine in Hordaland after 10,500 BP (possibly as young as 10,000 BP). Thus there is no overlap at all in the time of formation. Even the Ra on Sørlandet is probably younger than the Ra in Oslofjord. In Nordfjord the Nor moraine is of the same age as the Herdla moraine, while the

moraine in Trøndelag is of the same age as the Ra in Oslofjord.

ACKNOWLEDGEMENTS

The diagrams from different areas have been discussed with Prof. B.E. Berglund, cand. real.R. Sørensen, Prof. B.G. Andersen, cand. real. K.Anudsen, cand. real. J.L. Sollid, cand. real. A. Kjemperud and Prof. T.O. Vorren. Dr. O. Liestøl, Prof. B.G. Andersen, Prof. T.O. Vorren and Prof. H.E. Wright read through the manuscript critically. Dr. J.J. Lowe and Dr. J.M. Gray corrected the English language. To all these colleagues I proffer my sincere thanks.

Palaeoclimatic Inferences from Loch Lomond Advance Glaciers

J. B. Sissons

(University of Edinburgh)

ABSTRACT

Equilibrium firn line altitudes and relative dimensions of Loch Lomond Advance glaciers in Scotland and the Lake District indicate that snowfall was associated mainly with S to SE air streams preceding warm and occluded fronts, although SW air streams were more common. Precipitation was high in the western Grampians and in much of the ground west of the Great Glen but relatively low in the NW Highlands and very low in the NW Cairngorms and Speyside. This pattern is related to many depressions having followed tracks far to the south of those prevailing today, this in turn being related to the position of the oceanic polar front. The stadial evidence leads to a suggestion about the nature of ice-sheet growth.

There were pronounced variations in the distribution, size and altitude of glaciers that existed in Scotland and the Lake District during the Loch Lomond Stadial. For example, much of the SW Grampians along with the mountains W of the Great Glen were ice covered, whereas glaciers were mostly small in NW Scotland and the northern Grampians, being almost absent from the extensive Monadhliath Mountains. In the Lake District the largest glaciers developed along the W-E mountain axis and glaciers were far smaller in the N and NW of the area. Glacier firn lines below 300 m in part of the SW Grampians contrast with firn lines of 1000 m in the northern Cairngorms. These and other differences, along with glacier aspect, permit inferences to be made about the climate of the stadial.

Fig. 1, showing the limit of the Loch Lomond Advance in Scotland, is based on published work (Sissons, 1974a, 1977a, 1977b, 1977c, 1979a, 1979b; Sissons and Grant, 1972; Sissons and Sutherland, 1976; Gray and Brooks, 1972; Ballantyne and Wain-Hobson, 1979) and on unpublished work by R. Cornish, D. Rae, M. Robinson, D.G. Sutherland, K.S.R. Thompson, P. Thorp, T. Wain-Hobson and the writer. It includes the writer's corrections of his own mapping in part of the Ben Eighe Nature Reserve in NW Scotland (Sissons, 1977b, Fig. 9). The scale of Fig. 1 does not allow nunataks, which interrupted the larger areas of ice, to be shown. A broken line is used where the limit of the advance in the western Highlands has not yet been established.

The very large area of ice in the western Highlands, extending from Loch Lomond to the latitude of Skye, might seem to suggest that the principal snow-bearing air streams were from W or SW. However, evidence from many localities points to a different conclusion. For example, glaciers flowed out radially from corries in

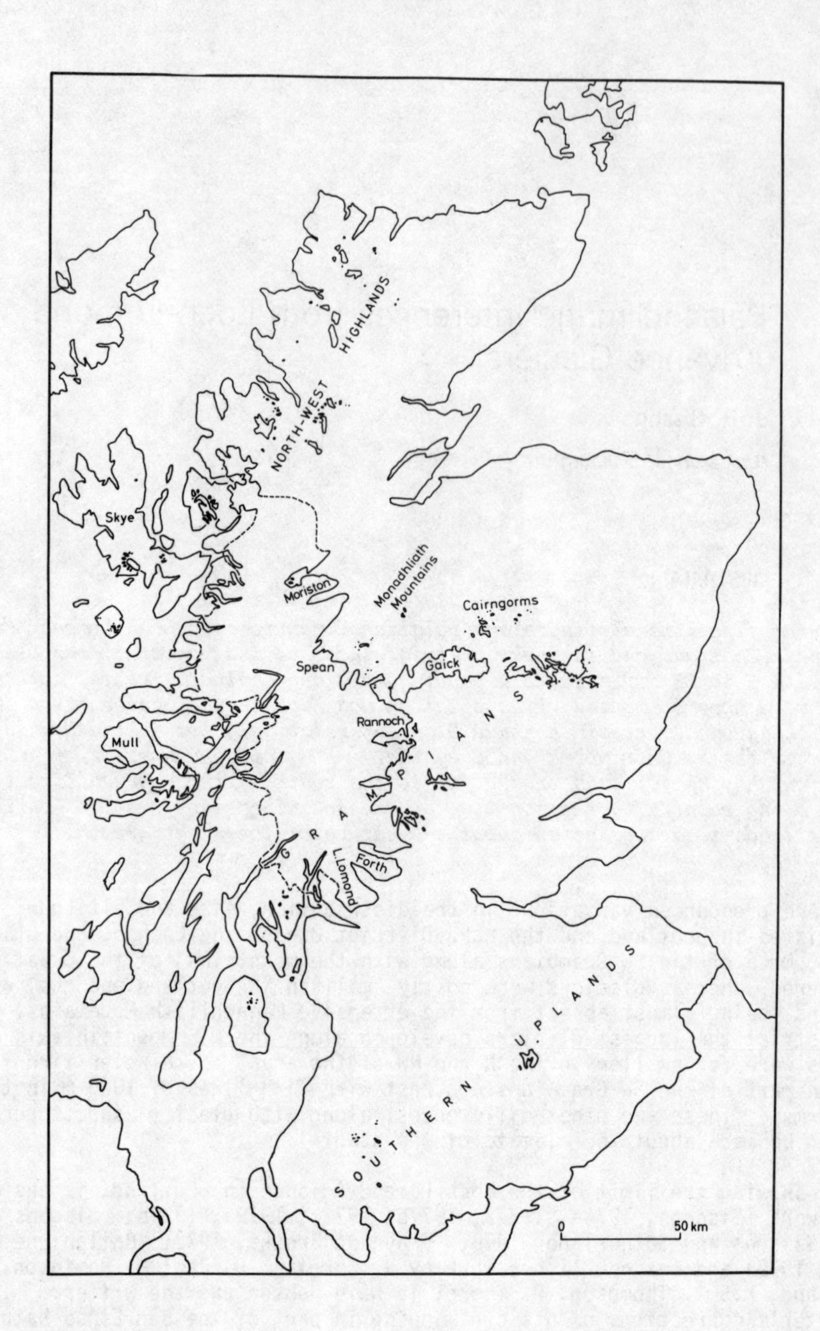

Fig. 1 Limit of the Loch Lomond Advance in Scotland.

the Cuillin Hills of Skye towards directions varying from SSW through W to NE, the largest flowing W and NW. It might reasonably be expected that glaciers occupying N- and NE-facing sites would have been larger and lower than those facing in opposite directions, owing to the effect of direct insolation. However, the combined volume of the two glaciers that faced SSW was three times that of the two glaciers that occupied N- and NE-facing corries, while the firn lines of the latter were higher by 300 m.

A plateau ice-cap existed in the central Grampians, its outlet glaciers occupying the upper parts of many valleys, including those of the rivers Truim, Garry and Bruar (Fig. 2). The greater part of the ice lay S of the axis of the high ground and the firn lines of the six portions of the ice-cap distinguished by the thicker broken lines in Fig. 2 rose from S to N. In the western Cairngorms firn line altitudes in part reflect local influences. Overall, however, they rose northwards, despite the glaciers in the northern part of the area having faced northwards (Fig. 3). The latter were also much smaller than some glaciers farther S that were adversely located in relation to direct insolation.

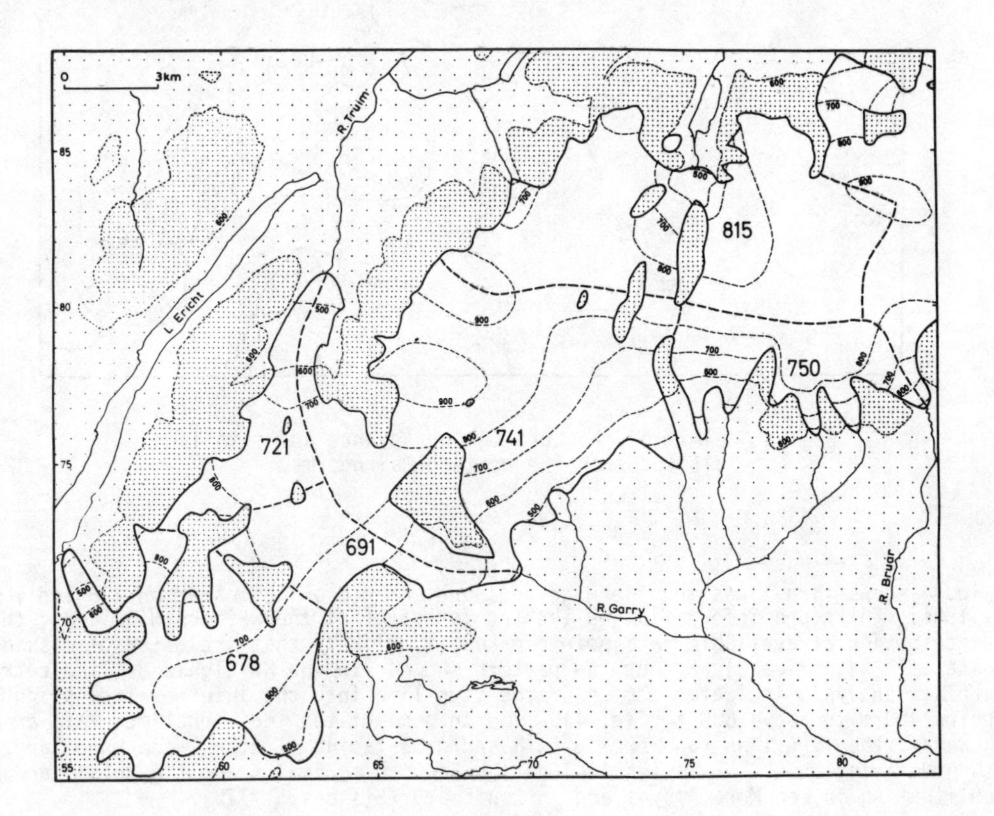

Fig. 2 Limit of the Loch Lomond Advance and firn line altitudes in and around the Pass of Drumochter.

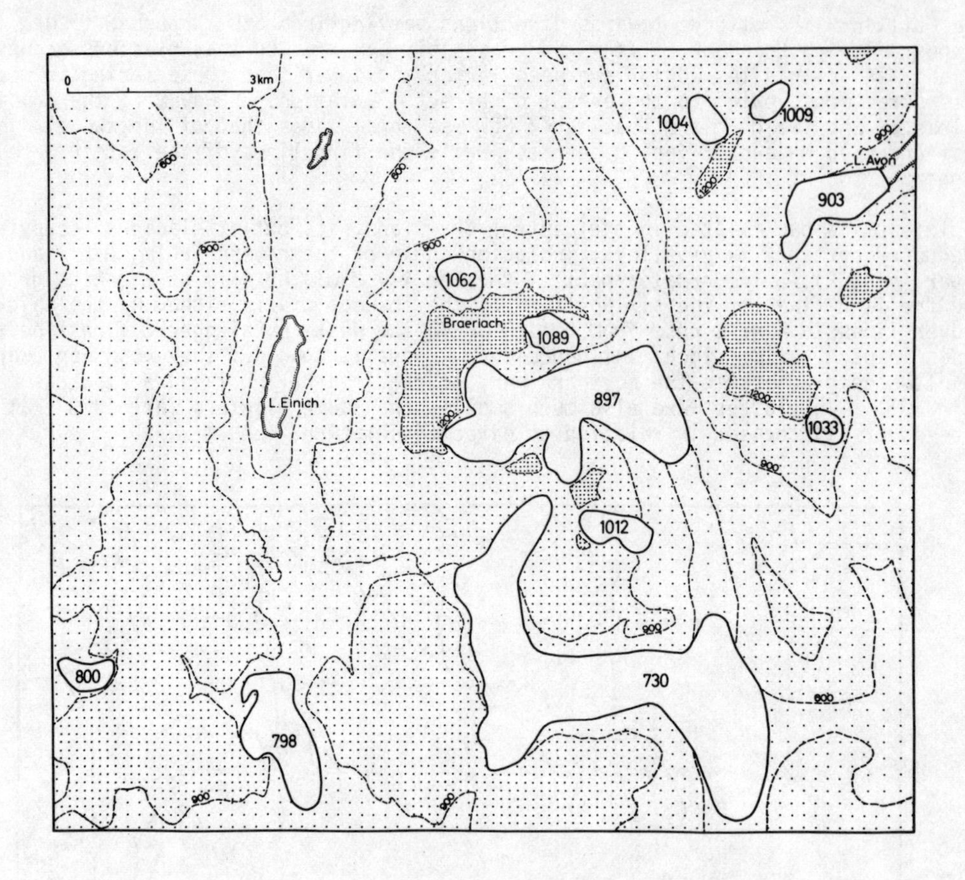

Fig. 3 Limit of the Loch Lomond Advance and firn line altitudes in the western Cairngorms.

Fig. 4 shows firn lines for the Highlands and Inner Hebrides based on 226 individual values. Although information is lacking for parts of the western Highlands, the distribution of available data points makes it unlikely that inclusion of missing data will significantly affect the pattern shown. In the NW Highlands linear trend surface analysis indicates a rise of the firn line into the interior from slightly below 400 m to above 600 m (Fig. 4). In some of the individual mountain areas, however, the firn line rose from S to N and/or S-facing glaciers were much larger than N-facing ones. Examples include the Fannich mountains and the upland areas culminating in Ben More Assynt and in Ben Dearg (Sissons, 1977b).

All the N-S differences in glaciers mentioned above occur in areas whose individual extent in this direction is no more than 25 km. Hence these differences cannot be attributed to latitudinal temperature differences. In fact, N-facing glaciers, being at higher altitudes, would have been favoured by lower temperatures, thus reducing ablation and also causing a higher proportion of annual precipitation to fall as snow. The differences between N- and S-facing glaciers would also have been strongly opposed by the influence of direct insolation. The only way in which the pattern can be explained is by invoking much heavier snowfall on the southern

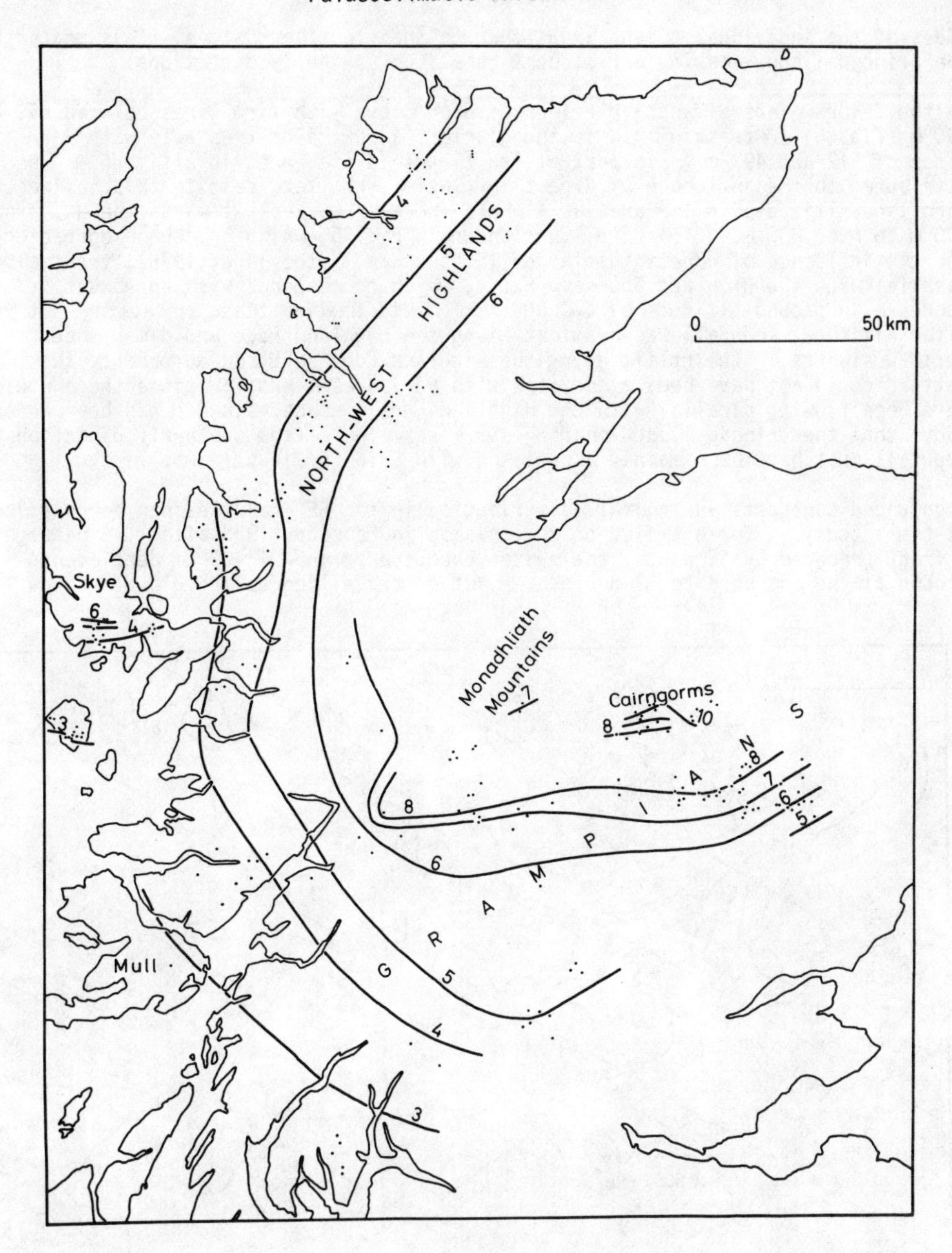

Fig. 4 Regional firn line altitudes (hundreds of metres) in the Highlands and Inner Hebrides.

sides of the individual upland areas than on their northern sides. This means that the principal snow-bearing air streams came from southerly directions.

In the SE Grampians S-facing glaciers by Glen Clova with firn lines between 660 and 690 m (Fig. 5) contrast with N-facing glaciers a few kilometres away with firn lines of 512 and 497 m. In part at least these differences in altitude may be attributed to the influence of direct insolation. Yet, despite this influence, firn line altitudes in the area as a whole increase markedly towards the NW, from 500 m to over 850 m. The firn lines shown in Fig. 5, which have been corrected for the influence of direct insolation (Sissons and Sutherland, 1976), trend almost parallel with the Highland Boundary Fault, which corresponds with an abrupt increase in ground altitude of 400-600 m. This implies that, on average at any given altitude, snowfall was heaviest along the Highland Edge and diminished thence north-westwards. The uplift along the Highland Edge required to produce this pattern could not have been associated with air streams from SW since these would have been flowing parallel with the Highland Edge. Hence, since it has been inferred above that the principal snow-bearing air streams were from southerly directions, snowfall must have been mainly associated with S to SE air streams.

Pronounced contrasts in snowfall distribution in the SE Grampians can be observed at times today. For example, on an occasion in February 1974 after the passage of a front preceded by SE winds, the writer encountered snow 15-20 cm deep even on low ground at the entrance to Glen Clova. Yet on travelling up the glen the snow

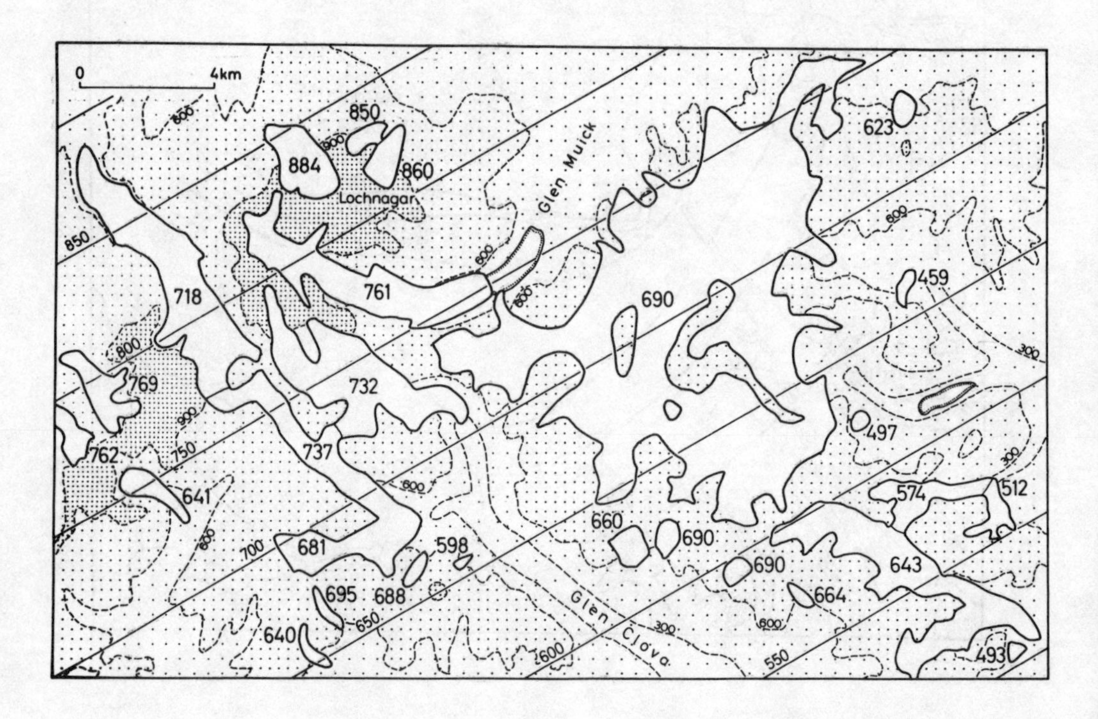

Fig. 5 The limit of the Loch Lomond Advance and firn line altitudes in the South-East Grampians. The linear trend surface of firn line altitudes corrected for the influence of direct insolation is also shown.

became thinner and, 15 km into the Highlands, was absent from the valley floor, while the adjacent hills and those to the north, including Lochnagar (1154 m), had only a sprinkling of snow. The rise in the firn line from 500 m at the Highland Edge to 1000 m in the northern Cairngorms shown in Fig. 4 is frequently paralleled by the snow line today during and after a spell of SE winds in winter or early spring. A typical situation along the Perth-Inverness (A9) road is for the snow line to be at about 100 m near the Highland Edge between Perth and Dunkeld, thence rising northwards to lie between 500 and 600 m on the northern slopes of the Cairngorms.

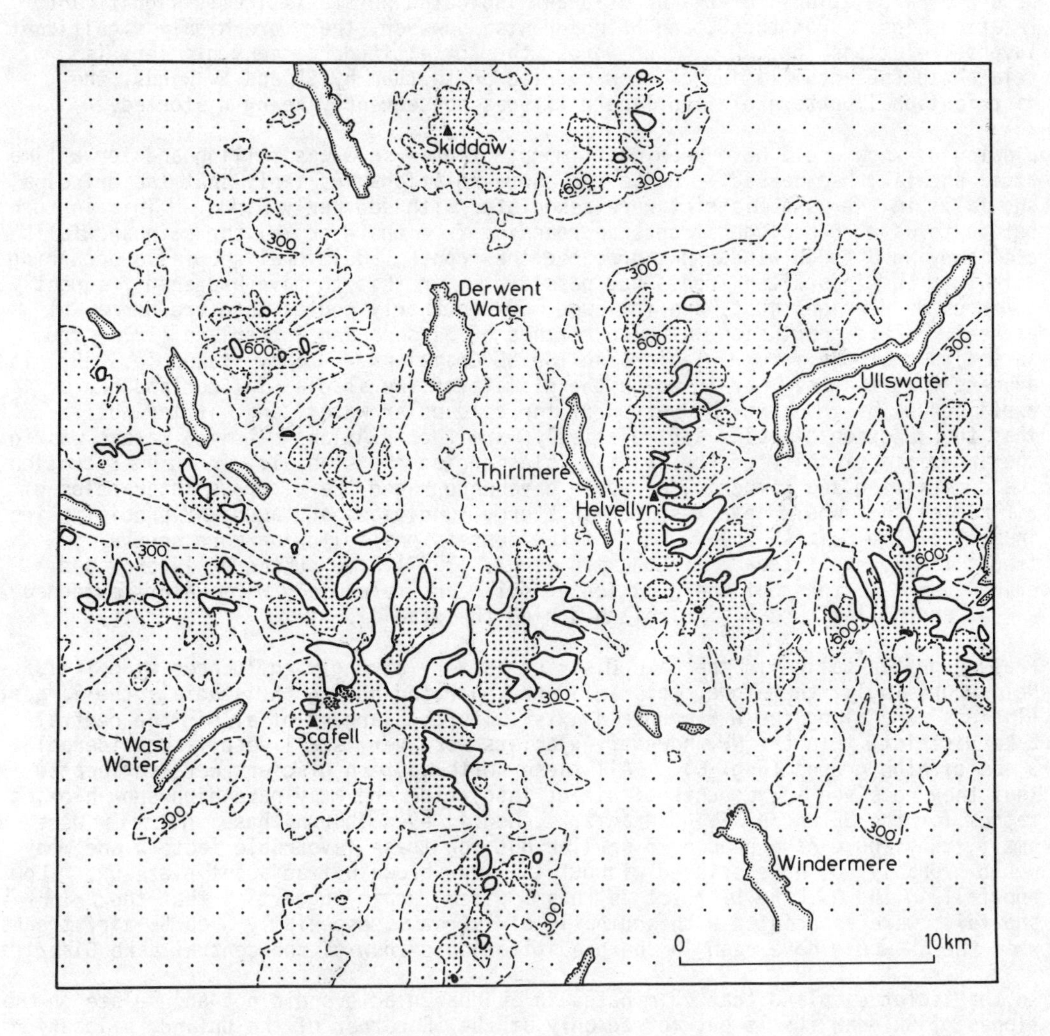

Fig. 6 The limit of the Loch Lomond Advance in the Lake District.

In contrast to firn lines in parts of the Scottish Highlands, those for the 64 Loch Lomond Advance glaciers that have been mapped in the Lake District (Fig. 6) show no significant gradient in any direction (Sissons, in press). An important factor influencing the altitudes of many glaciers was wind transfer of snow from surrounding higher ground. Potential snow-blowing areas for each glacier for each quadrant were measured on the Ordnance Survey 1:25,000 map, using a method described elsewhere (Sissons and Sutherland, 1976). Each snow-blowing area was divided by its associated glacier area to give a snow-blowing ratio. Regression of firn line altitudes for the 64 glaciers on the snow-blowing ratios for the NW quadrant and (in a separate calculation) for the NE quadrant indicated no statistically significant relationships. For the SE and SW quadrants, however, there are highly significant inverse relationships. In other words, the low altitude of many glaciers is related to the accumulation of snow carried on to them by SE and SW winds, the distributional pattern of firn line altitudes consequently being distorted.

Blowing of snow would have been most important when snow was falling and for a time after snowfall had ceased. Hence the above relationships imply that the principal snowfalls in the Lake District were associated with southerly winds. This in turn points to warm or occluded fronts approaching from the W or SW, the main snowfalls accompanying S to SE winds that preceded the fronts, additional snowfalls occurring with S to SW winds after fronts had passed. For this to have happened frequently numerous depressions must have followed more southerly routes than are normal at present. This conclusion was also reached by Sissons and Sutherland (1976) from an analysis of the former glaciers in the SE Grampians. They stated (p. 345) "it appears that the only way in which the climate of the stadial in Scotland can be explained is by invoking the return of the cold polar water" and "it may well be that the junction of polar and relatively warm (North Atlantic Drift) waters was in the immediate vicinity of the British Isles during the stadial. In such a situation, the contrasting sea temperatures would have encouraged the vigorous interaction of air masses that would have resulted in stormy conditions and abundant precipitation in much of the British Isles, many of the depressions following more southerly tracks than prevail today." Ruddiman *et al.* (1977) have subsequently provided evidence indicating that the junction of polar and relatively warm waters extended as far as the latitude of SW Ireland during the stadial.

Present precipitation in the Lake District is very strongly influenced by relief. Not surprisingly, therefore, this influence operated during the stadial: the largest glaciers were along the W-E mountain axis, with the largest of all in the central Lake District. In the NW, however, glaciers were very small despite considerable areas of high ground (Fig. 6). All these north-western glaciers were so located that they received low amounts of direct insolation and many have high snow-blowing ratios for the SE and/or SW quadrants. Hence, even though these glaciers were small, they would have been even smaller but for these favourable factors and many would probably not have existed without the snow blown off adjacent plateaux. Low snowfall in the NW Lake District is thus implied, again suggesting that the principal snowfalls were associated with southerly air streams, especially S to SE air streams, when the NW would have been in the precipitation shadow of the central Lake District.

In the isolated upland that culminates in Skiddaw glaciers did not accumulate on the slopes of Skiddaw itself but formed only at the SE corner of the upland, a location where S to SE air streams would have been subjected to strong uplift (Fig. 6). Similarly glaciers existed only at the SE corner of the Monadhliath Mountains (isolated 700 m firn line in Fig. 4). In the SE Grampians the group of three well-developed corries that overlooks Glen Clova is remarkable in that it faces S. The SE Grampians is unique in being the only area in which glaciers that existed during the stadial were exposed to SE air streams that had passed over the North Sea. It is therefore suggested that during late spring and summer, when sea ice would have thawed, frequent SE (to E) air streams caused much greater cloudiness than typifies the area today. Extensive stratus cloud would have been favoured by low sea temperatures and by snow and ice on the uplands. (Permanently overcast conditions are not, of course, suggested, for, as noted already, direct insolation influenced firn line altitudes.)

Regression analysis of the firn line altitudes of the former Lake District glaciers on their SE snow-blowing factors indicates an inverse relationship significant at the 0.005 level, the significance level with SW being 0.001 (Sissons, in press). For former glaciers in the NW Highlands there is no significant relationship between firn line altitudes and SE snow-blowing ratios, but an inverse relationship with SW is significant at the 0.02 level. For former glaciers in the SE Grampians the inverse relationship between firn line altitudes and SE snow-blowing ratios is significant at the 0.05 level, but that with SW is significant at the 0.001 level. These consistent results from widely separated areas suggest that SW winds were more important than SE winds in blowing snow onto glaciers. Since it has been inferred that snowfall was associated mainly with SE winds it can be further suggested that SW winds prevailed for longer periods than SE ones because (i) snow blowing is most effective when snow is actually falling, (ii) the SE winds that preceded fronts would normally have been stronger than the SW winds that followed fronts. However, the importance of SW winds may be exaggerated by the fact that snow blown off high ground by them would have tended to come to rest in locations more sheltered from effective direct insolation than would snow blown by SE winds.

At the present day moist, mild SW air streams are often associated with light or moderate, but sometimes prolonged, precipitation in the western Highlands and part of the Lake District, but often produce only drizzle or no precipitation at all in the eastern Highlands. In stadial times a similar distribution of precipitation with SW air streams may be inferred but the contrasts between W and E would have been emphasised. Extensive ice and snow surfaces in the W would have increased precipitation there and, as the major western ice mass built up (to ice-shed altitudes of about 850-900 m over Rannoch Moor and vicinity and in the area W of the Great Glen) its bulk would have added to the orographic influence. Hence considerable snowfall is likely to have been provided by SW air streams in the western Highlands and part of the Lake District (probably mainly with wind directions from between 180° and 225° rather than from between 225° and 270° (Sissons, in press)). This may appear to weaken further the inference made above that SW winds were more frequent than SE ones. However, the strong inverse correlation between firn line altitudes and SW snow-blowing areas in the SE Grampians cannot be attributed mainly to wind transfer while snow was actually falling, since considerable snowfall with SW air streams appears improbable as a frequent occurrence in that area. Furthermore, in the eastern Cairngorms, where it appears even less likely, there is strong evidence that SW winds were of major importance in the nourishment of certain glaciers (Sissons, 1979a). Hence it is maintained that SW winds were more common than SE ones.

The generalised firn line pattern for the Highlands (Fig. 4) strongly resembles the map of corrie floor altitudes provided by Robinson *et al.* (1971). The similarity would probably have been greater but for the over-generalization introduced by their trend surface analysis: for example, the high altitude of some of the Cairngorm corries is not evident from their map. In part the similarity reflects the influence of pre-existing corries on glacier accumulation. But some glaciers did not accumulate in corries and many that initially did so subsequently extended so far beyond their corries that the latter had minimal influence on firn line altitudes. When a large ice-sheet existed many corries were completely ice covered and in some instances ice flow was transverse or opposite to that of corrie ice under conditions of limited ice cover. Hence the altitudinal distribution of corries, along with their excellent development in many parts of the Highlands, implies numerous periods of limited ice cover during the Quaternary when climatic conditions were similar to those that prevailed during the Loch Lomond Stadial. In turn this suggests many occasions when oceanographic conditions were similar to those that applied during the stadial.

Precipitation at present in the former glacier accumulation areas of the western Grampians is very heavy, typically being 3000-4000 mm. When the Loch Lomond Advance glaciers were accumulating it would have tended to be less owing to lower air temperatures causing saturated air to contain less moisture. On the other hand, vigorous interaction of competing air masses would have tended to increase precipitation. On the assumption that these tendencies roughly cancelled each other, an indication of former summer temperatures may be obtained using Liestøl's graph of accumulation against summer (May-September) temperature at the firn line for certain Norwegian glaciers (Sissons, 1979d). Assuming that 20-25% of total precipitation was summer rain (*cf.* Manley, 1959; Sissons and Sutherland, 1976) 4°C is equivalent to about 4000 mm of annual precipitation and 3°C to about 2700 mm. Hence a temperature of 3.5 ± 0.5°C appears to cover reasonable possibilities. Using the 400 m firn line and an increase in temperature of 0.6°C/100 m descent, this indicates a May-September average sea-level temperature of 6.0°C, equivalent to an average July temperature of 7.5°C. Since these temperatures relate to the firn lines of glaciers in equilibrium, when climate is likely to have ameliorated slightly, stadial temperature would have been slightly lower, probably about 7°C July. For the Lake District (Sissons, in press) the comparable figure is 7.5°C (which agrees with Manley's (1959) figure derived from the former glaciers) and for the SE Grampians 6°C, this latter figure relating to the proximity of the North Sea and extensive summer cloud (Sissons and Sutherland, 1976).

The existence of permafrost down to sea-level in Scotland during the stadial is indicated by fossil frost wedges (Sissons, 1974b). Since the development of permafrost requires a mean annual temperature no higher than -1°C, the 7°C July value suggested for part of the western Grampians implies a January temperature no higher than -9°C. The mean annual temperature range of at least 16°C thus indicated is (at least) 50% greater than the present range. Winter temperatures on high ground must have been very low. Assuming a decrease in temperature of 0.6°C/100 m, a mean January temperature no higher than -17°C is indicated on the summit of Ben Nevis.

As Fig. 4 shows, there was a pronounced rise in the regional firn line from below 300 m in the SW Grampians to over 1000 m in the northern Cairngorms. That this rise was due essentially to differences in precipitation rather than to differences in temperature is indicated by the great contrasts in glacier dimensions between the two areas. In the SW Grampians the Loch Lomond glacier alone had a volume of about 80 km^3. Yet the 17 glaciers that developed in the Cairngorms had a combined volume of only 1.7 km^3. In the NW Cairngorms four corries at 1000 m altitude failed to nourish glaciers and the three glaciers that did exist in corries at this altitude were tiny. This implies very low precipitation here.

One way of estimating the precipitation is as follows. The average net accumulation on a glacier may be roughly calculated from its area above the equilibrium firn line and from its volume, if an assumption is made about the period involved. Using a period of 750 years, the value for the Loch Lomond glacier is slightly over 600 mm a year (water equivalent) and for the three N-facing corrie glaciers of the NW Cairngorms it is 100-150 mm a year. This suggests that the former received 4 to 6 times as much precipitation as the latter. Even using the suggested maximal value of 4000 mm precipitation for the SW Grampians, only 800 mm in the NW Cairngorms is indicated. Even this value appears too high, however, for several reasons: (i) the Loch Lomond glacier had the disadvantage that during much of its advance its terminus was in water, often deep, and would thus have suffered loss by calving; (ii) this glacier flowed southwards whereas the three Cairngorm glaciers flowed northwards; (iii) the latter would have benefited by being backed by plateaux from which snow would have been blown onto them by southerly winds; (iv) the above estimate includes an allowance of 20-25% for summer rain, which seems reasonable for the SW Grampians but is almost certainly too high for the high-altitude glaciers of the NW Cairngorms; and (v) the much higher altitude of the latter as compared with the Loch Lomond glacier would have favoured a lower rate of ablation. Hence it seems that annual precipitation in the NW Cairngorms would have been considerably less than 800 mm: one may suggest 500-600 mm. Since these figures refer to an altitude of about 1000 m, it is difficult to envisage more than 200-300 mm on the low ground of Speyside.

Annual snowfall for the three corrie glaciers of the NW Cairngorms was also calculated using the equation given in Sissons and Sutherland (1976). The same values were used for the constants in the equation as in that publication and annual snowfall was converted to annual precipitation in the same way. The values derived for the latter are 590, 600 and 640 mm.

Low precipitation in the NW Cairngorms and Speyside would have been favoured by the more southerly tracks than at present of many depressions, by low air temperatures, by the elimination of the North Sea as a significant source of moisture for N to NE air streams when it was frozen over in winter, and by the presence of the large ice mass in the western Highlands (once it had developed). The relatively low precipitation in the Cairngorms as a whole (along with the granite bedrock and altitude) was doubtless a major factor in the extensive development of excellent (now fossil) periglacial landforms such as lobes and sheets.

Birks and Mathewes (1978, p.469) inferred from the abundance of *Artemisia* pollen in the stadial deposits at a site in Speyside that the climate there was "rather arid" with warm summers. Low precipitation might reasonably be expected to have accorded with considerable amounts of sunshine, whose effectiveness would have been increased by astronomical factors in summer (Vernekar, 1971), thus leading to quite high day temperatures at this time of year. On the other hand, although there would have been some amelioration of climate in winter owing to the föhn effect, winters would probably have been extremely cold. Sea ice would have excluded the ameliorating influence of the sea on northerly air streams and, along with an ice-sheet over Scandinavia, would have favoured more frequent anticyclonic conditions than at present to the E and NE of Scotland, at times extending over Scotland and affecting especially the NE parts of the country. Such conditions, along with lower precipitation than now, and hence presumably less cloud, would have led to very low temperatures during long winter nights. Local contributory factors in Speyside would have been the topographic barrier of the major E-W Highland watershed lying to the S, cold air drainage (including glacier winds) from surrounding mountains covered with snow and ice, and the fact that the middle Spey valley is a major "frost hollow". It thus appears that the annual temperature range of at least 16°C suggested for part of the western Grampians would have been considerably exceeded in Speyside.

Interpreted as mainly controlled by snowfall, the regional firn line pattern in the Highlands (Fig. 4) accords with the distribution of *Artemisia* as revealed by pollen analyses. Species of *Artemisia* are associated with dry ground and normally do not tolerate much snow cover. Expressed as a percentage of total land pollen, *Artemisia* fails to attain 10% in stadial samples from Loch Sionascaig near the coast of NW Scotland but it increases inland and at Loch Craggie, east of the main N-S watershed, it achieves 25%. At Loch Tarff by the southern end of Loch Ness the maximum is 30%. Near the junction of the Highlands with the Central Lowlands maximal *Artemisia* percentages range from about 5 to about 20 but near the head of the Spey valley (Loch Etteridge) they reach 30-40. Farther down the valley (Abernethy Forest) the maximum is 65% and values between 40 and 65% are maintained until almost the end of the stadial (Pennington *et al.*, 1972; Walker, 1975a, 1975b, 1977; Lowe, 1978; Lowe and Walker, 1977; Birks and Mathewes, 1978).

Fig. 1 shows that the area covered by glaciers in the NW Highlands was very small. This can only in part be attributed to the smaller area of high ground compared with more southern parts of the Highlands, for the regional firn line does not descend as low in the NW as might be expected for the latitude: the regional firn line in the SW Grampians continued 100 m below that in the NW (Fig. 4). In the Lake District the average firn line altitude of the 64 glaciers is 540 m, while that of 68 glaciers in the NW Highlands is 520 m. Since the latter area is the more northerly by 400 km, lower summer temperatures may be expected to have prevailed, while glaciers in the NW Highlands would have been much favoured by the lower angles of incidence of direct insolation as compared with the Lake District. Hence some major factor adverse to glacier development must have been over-riding these favourable influences in the NW: the only factor that can be reasonably invoked is snowfall, which must have been relatively low. Snowfall in the NW Highlands was probably associated mainly with occasional depressions whose centres normally passed to the NW of the area as they tracked north-eastwards. The snowfall deficiency in the area may be attributed to the majority of depressions having followed southerly tracks, this in turn being related to the position of the polar front in the Atlantic Ocean.

Fig. 1 depicts an extremely small area of ice in the Southern Uplands: this is partly because some small glaciers are omitted since detailed information is not available, but in the western Southern Uplands (in the Loch Doon basin and surrounding hills) all the glaciers are shown (R. Cornish, pers. comm.). Yet during ice-sheet conditions this area had a major ice-shed from which ice flowed northwards before being deflected by Highland ice. Southern Uplands ice extended to within a short distance of the site of Edinburgh and its importance is demonstrated, for example, by the extensive and well-developed ice-moulded landforms of the Tweed basin. If Fig. 1 is regarded as portraying a stage in the build-up of an ice-sheet, it is difficult to envisage such a large ice mass developing on the Southern Uplands before they were overwhelmed by Highland ice. One might invoke a general lowering of the firn line, which would have allowed considerable areas of the Southern Uplands to become glacier accumulation areas: yet this would also have greatly benefited the already extensive Highland ice. A solution to this problem is suggested by the snowfall deficiency in the NW Highlands during the Loch Lomond Stadial, along with low snowfall in the NW Cairngorms and Speyside, as well as in the Monadhliath Mountains (where glaciers existed only in the extreme SE). Since this large area of low or relatively low snowfall appears to be related to the more southerly tracks than at present of many depressions, it is suggested that, as an ice-sheet developed and the oceanic polar front moved southwards beyond its stadial limit, a zone of maximal snowfall also moved southwards, with relatively dry conditions following in its wake. Thus the Southern Uplands would have benefited at the expense of the Highlands. Similarly, one may account for the powerful ice dome that developed over the Lake District, contrasting with the mere 3.3 km^3 of ice that accumulated during the stadial. Subsequently the Welsh uplands, which had only tiny glaciers during the stadial but a large plateau ice-cap feeding into the ice-sheet, would have benefited.

The above suggestions serve as a reminder that the whole of the discussion in this paper has to be qualified by the fact that, during the Loch Lomond Stadial itself, the oceanic polar front was not stationary: the former glaciers, as considered above, only permit inferences about average climatic conditions over a period of varying duration. Ruddiman *et al.* (1977) suggested that the oceanic polar front moved southwards at an average rate of 1500 m/yr and that retreat averaged 1650 m/yr. Such figures can only be regarded as crude initial approximations, however, and it is not yet clear whether or not the polar front remained at or near its southern limit for some time.

Acknowledgements

The writer is grateful to C.K. Ballantyne, R. Cornish and M. Dale for comments on the draft of this paper.

Shore Erosion by Frost: An Example from the Scottish Lateglacial

A. G. Dawson

(Lanchester Polytechnic)

ABSTRACT

A well-developed raised shore platform (the Main Rock Platform) is described from Jura, Scarba and NE Islay, Scottish Inner Hebrides. Levelling of 175 platform fragments (approximately 1,000 altitude measurements) indicates that the feature declines in altitude to the SW with a regional gradient of 0.13 m/km. It is suggested that the feature was formed by periglacial shore erosion during the Loch Lomond Stadial (Younger Dryas) and is of the same age as the Norwegian Main Line (P12). It is also proposed that the traditional correlation between the Main Rock Platform and a low Irish (interglacial) platform should be abandoned. Processes of shore platform development in polar areas are also discussed.

INTRODUCTION

In the Scottish Inner Hebrides a raised shore platform in Dalradian quartzite is developed along the coast of Jura, Scarba and NE Islay (Fig. 1; Plate 1). The feature was originally considered by Wright (1928, p.99) as having been formed by the postglacial sea that he called "... the cliffmaker *par excellence* ...". Later McCallien (1937) suggested that since modern platform development on hard rock coasts of the British Isles was minimal it was unlikely that the postglacial sea could have formed such a wide platform. McCallien proposed instead that the feature is interglacial in age, a view later endorsed by Stephens (1957), Donner (1963), Synge (1966) and McCann (1966). The platform was correlated with a low till-covered interglacial platform in Ireland described by Wright (1937), Movius (1942) and Stephens (1957). Later Gray (1974) described a well-developed platform along the Firth of Lorn and eastern Mull and showed that it declines in altitude W and SW with a gradient of 0.16 m/km. Gray (1974) considered that the platform (referred to by him as the Main Rock Platform) is interglacial in age and attributed its regional altitude variation to later tectonic subsidence.

Sissons (1974c) showed that in SE Scotland a buried erosional shoreline (the Buried Gravel Layer) declines in altitude eastward with a gradient of 0.17 m/km and argued on several grounds that this shoreline is of the same age as the Main Rock Platform of W Scotland, both having been formed during the cold climate of the Loch Lomond Stadial (Younger Dryas). Sissons (1974c) suggested that McCann's (1966) evidence of glacial modification of the platform was unconvincing and believed instead that, since delicate stacks and arches occur on the platform surface in areas where

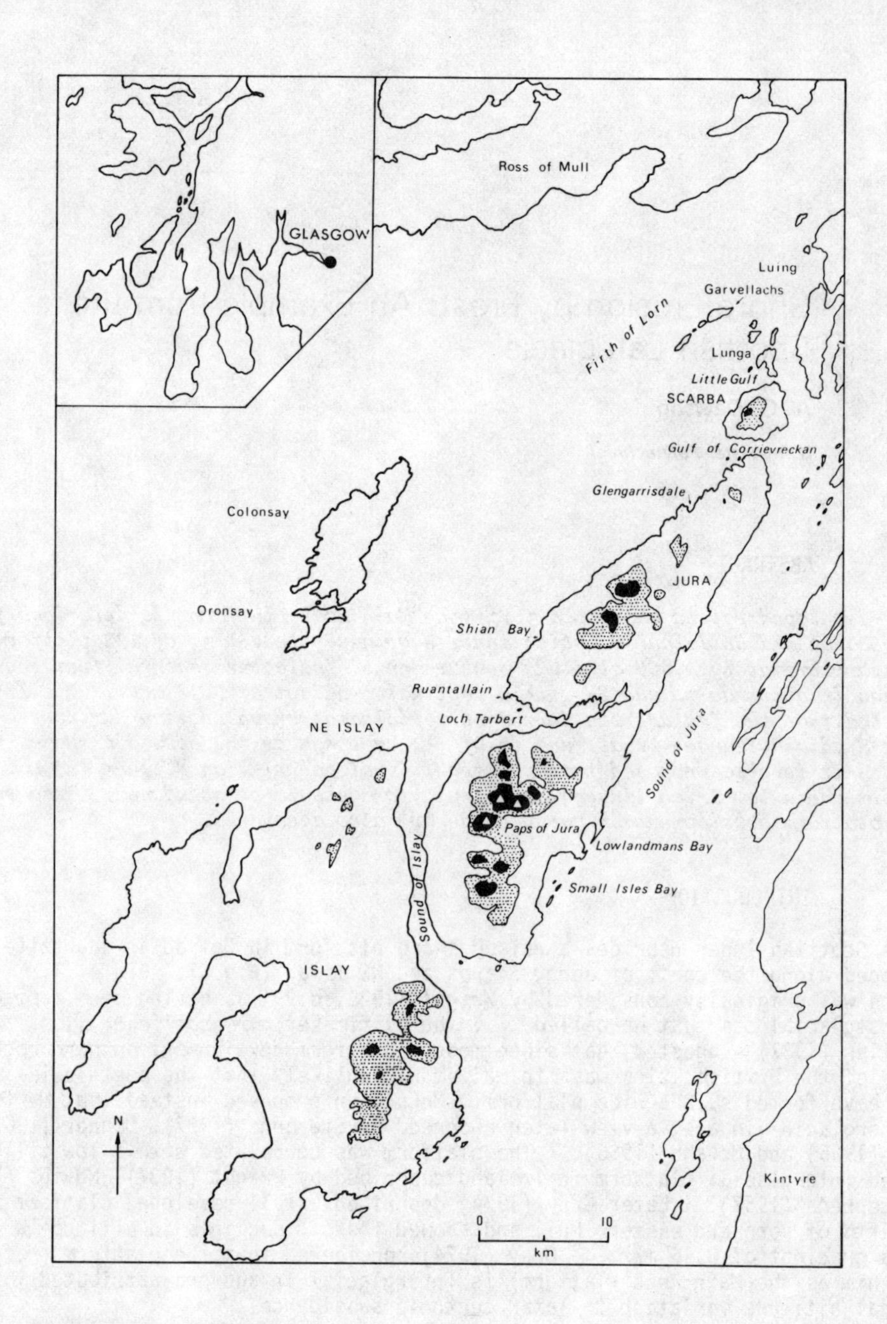

Fig. 1 Location of Jura, Scarba and NE Islay. Ground areas over 200 m (shaded) and 400 m (black) are also shown.

Plate 1 View (looking southwest) of Main Rock Platform, NW Jura.

glacial erosion was formerly intense, the platform is unlikely to be of an interglacial age. Later Gray (1978) measured the Main Rock Platform in Knapdale, Kintyre and neighbouring areas and found that it declines in altitude southwards with a gradient of 0.12 m/km. Gray abandoned the interglacial hypothesis and concluded that the feature is Lateglacial in age.

MORPHOLOGY

The platform of Jura, Scarba and NE Islay is the SW continuation of that described farther north and east by Gray (1974, 1978) and is thus referred to here as the Main Rock Platform. However in SW Jura, the Main Rock Platform crosses a low intertidal rock platform (the Low Rock Platform) that passes beneath thick accumulations of glacial till (Dawson, 1979, Plate 5). The Low Rock Platform surface is glacially striated and is characterised by numerous pot-holes and water-rounded rock. This second shore platform varies in altitude between 1 and 2 m OD and is horizontal throughout Jura, Scarba, NE Islay, Colonsay and Oronsay (Dawson, 1979, pp.132-42), the feature being separated from the Main Rock Platform by a small cliff generally 1 - 4 m in height. In contrast the Main Rock Platform varies regionally in altitude and possesses a number of morphological characteristics that are common to all localities where platform fragments occur:

1. Platform fragments are often exceptionally well developed in areas where the maximum fetch is only several hundred metres. Good examples of such fragments are north of Glengarrisdale, NE Jura (NM 685004); along the northern coast of Scarba; Lowlandmans Bay, SE Jura (NR 562720); and along the sheltered coastline of Loch Tarbert (Fig. 1).

2. The platform fragment surfaces, although generally horizontal, are often irregular in detail and do not possess the well-developed ramp abrasion profiles characteristic of platforms formed by marine abrasion (*cf*. Bradley, 1958; Hills, 1972).

3. Numerous raised sea-stacks and natural arches protrude above the platform while pot-holes and water-rounded rock surfaces are rarely present. In addition the cliff backing the platform is often indented by numerous geos and caves: in the study area 97 raised caves have been identified.

4. The platform surface has numerous crescentic concussion scars that have resulted from the former beating of shingle on the rock surfaces.

5. There is no evidence of glaciation of the Main Rock Platform.

SHORELINE MEASUREMENT AND ANALYSIS

In the study area, Ordnance Survey bench marks were used as a datum and the altitudes of all Main Rock Platform fragments were determined using Hilger Watts and Wild NAK autoset levels and a staff graduated to 0.01 m. Since the Main Rock Platform is generally bare of overlying sediment, the altitude of its inner edge was obtained by direct levelling of bedrock. Due however to the irregular nature of the rock surface at the platform inner edge, the altitude of individual platform fragments was established by the levelling and averaging of 4 - 15 rock altitudes considered representative of the inner edge at each site selected for measurement. In this way the altitudes of 175 fragments were derived from approximately 1,000 measurements.

The altitudes of all platform fragments were plotted on five shoreline height-distance diagrams in the quadrant between west and south from an arbitrary point of origin NE of Scarba (Dawson, 1979). Comparison of shoreline gradients calculated for the projections indicates that the NE-SW projection plane lies approximately normal to the isobases (Fig. 2). Owing to the regular decline in platform altitude on the diagram (Fig. 2) and the continuity of the feature in the field it is believed that all fragments are part of a single feature. From the shoreline diagram it can be seen that the platform declines in altitude to the SW from 7 m OD in N Scarba to sea-level in NE Islay and has a gradient of 0.13 m/km. The values are consistent with those measured in neighbouring areas farther north by Gray (1978, Fig. 5c) where the platform rises in altitude NE with a similar gradient from 7 m OD on Luing (immediately north of Scarba) to 10 m OD in the Oban area.

Measurement of Main Rock Platform width and cliff height at 526 regularly spaced locations in W Jura was undertaken in order to determine the amount of rock formerly removed from the coastal zone (Dawson, 1979). Since the Main Rock Platform merges with and crosses the Low Rock Platform in NE Islay and SW Jura these areas were excluded from the analysis, for it is likely here that the Main Rock Platform developed partly on the pre-existing Low Rock Platform surface. From these measurements the product of platform width and cliff height indicates that an average of 2,074 m^3 of rock per metre of coast was removed. If it is assumed instead that a uniform seaward slope was in existence prior to platform development, an average of 1,037 m^3 of rock per metre of coast was removed. This figure is equivalent to the removal of almost 50 million m^3 of rock from the W Jura coast. It should be noted however that the values of 2,074 m^3 and 1,037 m^3 of rock per metre of coast are mean volumes of rock removal and that locally much larger volumes of rock were removed during platform development. For example at Ruantallain (Fig. 1) as much as 9,800 m^3 of rock per metre of coast may have been removed (Dawson, 1979, p.321).

INTERPRETATION

Age

The interglacial hypothesis fails to explain the occurrence on the platform surface of delicate stacks and arches, the calculated shoreline gradient of 0.13 m/km away from the centre of glacio-isostatic uplift and the absence of evidence demonstrating glaciation of the platform. Although striated and ice-moulded platform surfaces have been reported in the Firth of Lorn area (McCann, 1966; Synge, 1966; Peacock, 1975b), Gray (1978) has noted that at the described locations the glacial evidence is on the frontal slope of the platform and not on its surface. Moreover, striae occurring on the platform surface need not indicate over-riding by glacier ice. For example, Dionne (1973, p.185) and Gray (1978) have suggested that striated platform surfaces may result from the grounding of pack-ice or small ice-bergs. In addition, the presence of till overlying platform surfaces (McCann, 1966, p.89) can result from the slumping or solifluction of till from backing cliffs and need not imply glaciation of the platform (Gray, 1978, p.159). Finally, the writer believes that many of the ice-moulded platform surfaces that have been described by McCann (1966) and Synge (1966) are Low Rock Platform fragments. Many of the coastal locations where ice-moulded platforms have been described are in areas where the sloping Main Rock Platform crosses the Low Rock Platform (*e.g.* in Colonsay and Oronsay (McCann, 1968)).

The regional gradient of the feature is extremely difficult to explain unless the feature is interpreted as having been formed since the decay of the last (Devensian) ice-sheet. Since the gradient of the highest Lateglacial shoreline in Jura and NE Islay is 0.61 m/km and that of the Main Postglacial Shoreline is 0.05 m/km (Dawson, 1979) it is likely that the platform was produced during part of the intervening period. These figures are consistent with formation of the feature during the Loch Lomond Stadial, being similar to gradients calculated for areas farther north and east by Gray (1974, 1978) (0.12 m/km and 0.16 m/km) and for SE Scotland by

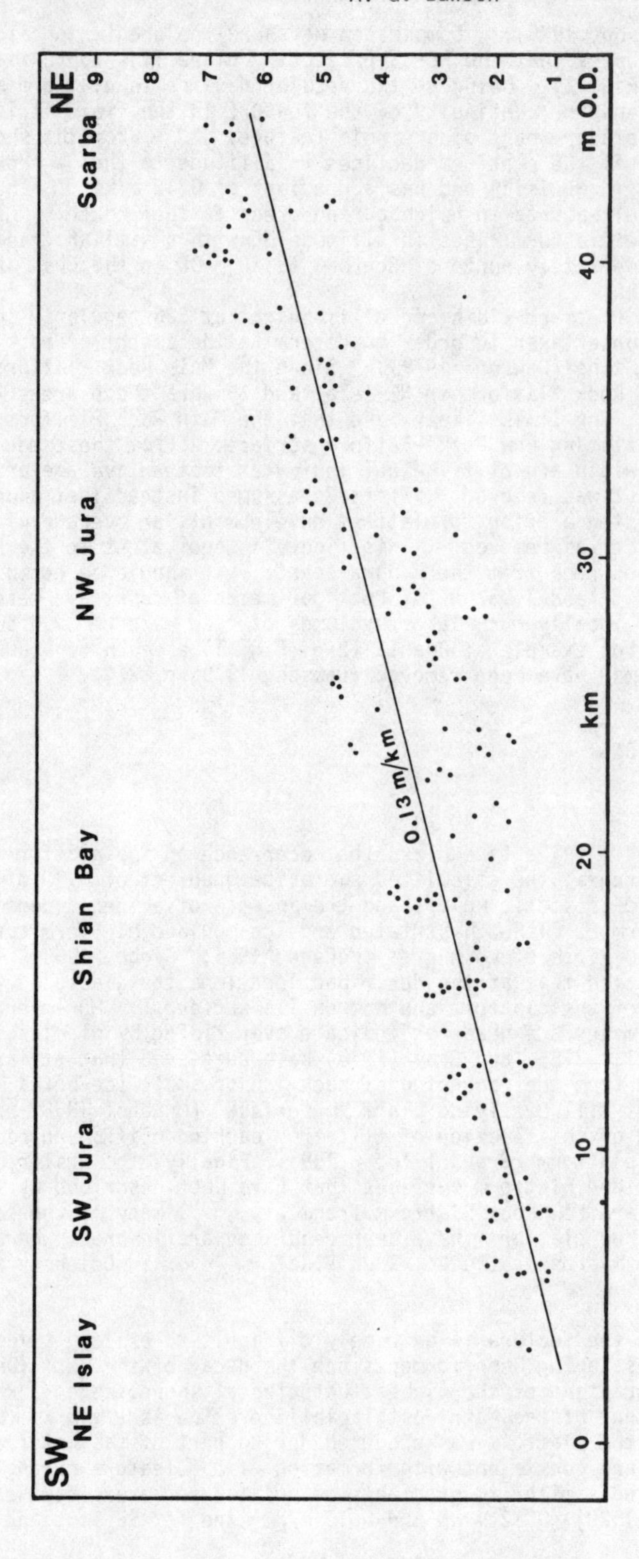

Fig. 2 Linear regression of Main Rock Platform fragment altitudes

Sissons (1974c) (0.17 m/km).

Origin

On the above interpretation a difficulty that appears to arise is the production of such a well-developed shoreline during the relatively short period of the Loch Lomond Stadial. Since modern marine erosion of rock is negligible in areas where the Main Rock Platform is best developed it appears that during the Loch Lomond Stadial, coastal erosion was capable of rapidly cutting the "... wide shelves in hard rocks in sheltered waters ..." described by Steers (1952, p.183). Assuming a duration of 1,000 years for the length of the Loch Lomond Stadial, it can be calculated from the data given earlier that in this area, if a uniform seaward slope was in existence prior to platform development, the average rate of rock removal was 1.04 m^3 per metre of coast per year: equivalent to a cliff retreat rate of 7 cm/year. This figure may be compared with a maximum rate of cliff retreat of 0.9 cm/year above a fossil rock glacier in Jura interpreted as also having formed during this period (Dawson, 1977) and with a value of 1.7 cm/year (Sissons, 1976a, p.189) for cliff retreat above a fossil protalus rampart in Wester Ross. The above figures for inland cliff retreat are not directly comparable however, since unlike coastal cliffs, the cliff face areas behind the fossil rock glacier and protalus rampart were gradually buried by talus aggradation. The calculated rate of cliff retreat of 7 cm/year may also be compared with retreat rates of 2.5 - 5.0 cm/year measured by Jahn (1961, p.22) on a modern coastal cliff in Spitzbergen.

It therefore appears that the formation of the Main Rock Platform during the cold climate of the Loch Lomond Stadial demands very special conditions in order to explain its origin in such a comparatively short period of time. Sissons (1974c) and Gray (1978) suggested that these conditions were characterised by processes of frost-shattering of rock in the semi-diurnally wetted intertidal zone and cliff base. Gray (1978) added that marine abrasion and debris removal may have been assisted by the apparent storminess during the Loch Lomond Stadial (Sissons and Sutherland, 1976). In addition he proposed that since diurnal temperature fluctuations in W Scotland during the Loch Lomond Stadial were most probably greater than in high latitude coastal areas, very intense frost-shattering of rock in the coastal zone would have favoured the development of wide platform fragments.

Additional evidence for the development of shore platforms by periglacial shore erosion during the Loch Lomond Stadial is provided along the shores of the former ice-dammed lake of Glen Roy in the Scottish Highlands (Sissons, 1978). Sissons concluded that during the brief existence of the lake, shore platforms reaching widths of 10 m or more were cut in metamorphic rocks at several levels. The existence in Glen Roy of well-developed shore platforms associated with former ice-dammed lakes (also areas of restricted fetch) further indicates that shore platform development occurred at extremely rapid rates in Scotland during the Loch Lomond Stadial.

Processes

The efficacy of shore erosion of rock on modern polar coasts is the subject of disagreement. Although several authors have suggested that polar shore platform development is slow (*cf.* Zenkovich, 1967; Bird, 1967; Davies, 1972; French, 1976), others have described recently eroded platforms (*e.g.* Jahn, 1961; Araya and Hervé, 1972; Calkin and Nichols, 1972; Sollid *et al.*, 1973; Moign, 1973, 1974). These studies have shown that polar shore platforms differ in morphology from non-polar shore platforms and are formed by different shore processes. Non-polar shore platforms are primarily the result of hydraulic action and corrasion during storms (King, 1972; Komar, 1976) and are often characterised by inclined ramp abrasion profiles. Wentworth (1938) suggested that although ramp abrasion and wave quarrying (scour) are important factors in non-polar platform development, water-layer weathering is also important and results in the later planation of platforms. In all previous studies however, modern non-polar platforms are said to be best developed in areas of exposed fetch and are poorly developed in areas of restricted fetch.

Owing to the important role of storm waves in the development of non-polar platforms it has been considered by many workers that, since storm wave activity in polar areas is restricted by the seasonal presence of pack ice and the ice-foot, wave processes in polar areas accomplish little erosion. However, Rekstad (1915), Vogt (1918) and Nansen (1922) have shown that shore erosion by frost action (rather than wave action) is an important mechanism of platform development in polar areas. Nansen (1922), for example, observed the development of an ice-foot on coastal rock ledges and the accumulation of frost-rived debris at the foot of adjacent slopes. He proposed that in polar areas the function of waves was primarily to remove frost-rived debris to offshore areas and added that, since the riving capacity of freezing freshwater was greater than that of freezing sea-water (*cf.* Guilcher, 1958, pp.25-6), shore erosion by frost is most effective in areas where freshwater can freeze on rock surfaces (*e.g.* at the foot of cliffs where subaerial run-off results in the development of a freshwater ice-foot). Nansen's views were later reiterated by Grønlie (1924) and Holtedahl (1960), the latter observing (p.528) that " ... contemporary marine erosion in N Norway has truncated the more rounded coastal landforms ...". The above authors all noted that, owing to the important role played by frost-riving in causing cliff retreat, polar platforms are often exceptionally well-developed in areas sheltered from severe wave attack, a fact which led Guilcher (1958, p.69) to state "... there seems no doubt that the ledges which surround many Norwegian fjords at about 0.5 m above mean sea-level are due to coastal frost-shattering. They cannot be explained by abrasion in such sheltered waters with so limited a fetch."

It is therefore clear that coastal cliff retreat occurs rapidly in polar areas and results in the development of shore platforms in areas of both exposed and restricted fetch. In addition, frost-riving on intertidal rock surfaces results in the development of irregular platform surfaces. Moreover, owing to the limited role played by wave action in polar platform development, well-developed ramp abrasion profiles characteristic of platforms produced in non-polar areas are rarely present. It should be emphasised here however, that in certain high latitude coastal areas, polar and non-polar coastal processes occur together and result in the development of platforms of 'intermediate' morphology. For example on polar coasts where there is less inshore summer pack ice, storm wave activity may be a more important factor in platform development.

Previous studies have also revealed the influence of rock type on the development of polar shore platforms. Holtedahl (1929, pp.163-4) noted the efficacy with which lavas, tuffs, phyllites, schistose limestones and quartzites are eroded by polar shore erosion. Battle (*cf.* Bird, 1967) suggested that frost-riving is ineffectual in non-porous rocks such as granite but is a powerful denudation agent in more brittle porous rocks. Nansen (1922) and Hoel (1909) noted the susceptibility of gabbro, argillaceous schists and shales to polar shore erosion and observed that igneous dykes, granite and gneiss are particularly resistant to frost-riving. Kranck (1950, p.26) pointed out that polar shore platforms are often well-developed in granite and gabbro.

It is therefore suggested that the Main Rock Platform of Jura, Scarba and NE Islay is morphologically very similar to shore platforms presently being formed in polar areas. Moreover, the presence of wide platform fragments in quartzite need not occasion surprise since this rock is particularly susceptible to frost-riving.

Relation to the Norwegian Main Line

Sissons (1974c, p.47) proposed that the Main Rock Platform and the Buried Gravel Layer of Scotland were formed synchronously with the Norwegian Main Line (P12) described by Marthinussen (1960) in N Norway. Marthinussen (1960, p.418) noted that a characteristic of this shoreline "... which is sometimes more a zone than a line, is that commonly it has been abraded in rock often with a very marked rock terrace. Therefore it probably embraces a considerable period of time with only slight fluctuations of relative sea-level." Andersen (1968, p.138) also stated that, "... the terraces of the Main Line are thought to have been formed by more processes than marine abrasion alone such as frost shatter and erosion by sea-ice and bergs." Andersen also noted that the Main Line is developed irrespective of exposure and is commonly located in narrow fjords and sounds while Sollid *et al.*, (1973, p.245) argued that the clearly defined rock terraces of the Main Line show that it most likely belongs to a period of minimal shoreline displacement. Since rates of isostatic uplift in N Norway were considerably faster than those that occurred in W Scotland during the Loch Lomond Stadial (the Main Line gradient being 1.2 - 1.4 m/km (Aarseth and Mangerud, 1974, p.19)) the presence of the well-developed Main Line shore platform in Norwegian fjords indicates that there the rate of platform development was extremely rapid. The slower rate of glacio-isostatic uplift in W Scotland meant that periglacial shore erosion of rock was more concentrated than in Norway and resulted in the development of an extremely clear platform and cliff.

Relation to the Irish (interglacial) Platform

Since it has been shown that the Main Rock Platform passes beneath sea-level in SW Jura, NE Islay and Colonsay, the traditional correlation between the Main Rock Platform and the low rock platform in Ireland (Stephens, 1957) is erroneous.

Relation to England and Wales

The evidence cited here of rapid periglacial shore erosion suggests that similar erosion occurred along the coastline of England and Wales not only during the Loch Lomond Stadial but also during earlier stadials. Therefore raised or submerged shore platforms in these areas need not necessarily be considered as having formed during long interglacial periods of relative sea-level stability.

ACKNOWLEDGEMENTS

Financial assistance was provided by the University of Edinburgh and the Carnegie Trust for the Universities of Scotland. The writer is grateful to Dr. J.B. Sissons for commenting on the draft of this paper and to Mr. D. Gilmour for assistance during levelling.

Lateglacial Environmental Changes Interpreted from Fossil Coleoptera from St. Bees, Cumbria, NW England

G. R. Coope and M. J. Joachim

(University of Birmingham)

ABSTRACT

A diagram is presented showing the stratigraphical occurrences of 210 named species of Coleoptera in the Lateglacial section exposed in the sea cliff at St. Bees, Cumbria, NW England. A series of 11 radiocarbon dates from these deposits is also given. Using the present-day geographical distributions of each species of Coleoptera, it has been possible to divide the fossil assemblage into a series of faunal units, each typified by species with similar present-day ranges. The insect fauna thus shows that the Windermere (Lateglacial) Interstadial was at its warmest prior to 12,200 BP, and that the climate was then somewhat warmer, at least during the summer, than at the present time. The episode of climatic amelioration began well before the commencement of deposition of organic-rich sediments in this sequence. Deposits laid down during the 'Alleröd period' contain fossil Coleoptera that indicate cooler conditions. During classical pollen zone III times (the Loch Lomond Stadial) the insect fauna had definite arctic affinities and included a high proportion of obligate arctic/alpine species that are no longer found living in the British Isles. This sequence of climatic events interpreted from fossil Coleoptera is entirely in keeping with similar data from other Lateglacial sites in Britain and Ireland.

INTRODUCTION

The Quaternary deposits exposed in the cliffs SW of St. Bees (Fig. 1) on the Cumbrian coast have for long attracted the attention of geologists (Smith, 1912; Eastwood *et al.*, 1931; Trotter and Hollingworth, 1932). The palynology of the Lateglacial deposits was described by Walker (1956), who subdivided the pollen diagram into three noncommittal zones, α, β, and γ. These were later attributed to the classical pollen zones I, II and III respectively by Pearson (1962) during an investigation of the insect fauna from the same locality. Pearson established the richness of the fossil Coleoptera in the St. Bees deposits, but unfortunately it has not been possible, on the basis of the published data, to relate the occurrence of most of the species to the pollen zonation. In the light of the apparent discrepancies between the palaeoclimatic inferences drawn from the coleopteran succession and the classical interpretation of Lateglacial pollen diagrams (Coope, 1970a; Coope and Brophy, 1972; Osborne, 1972), it was decided, with Dr. Pearson's kind permission, to re-examine the fossil Coleoptera from this section.

The Lateglacial deposits take the form of a sequence of clays, sands and detritus

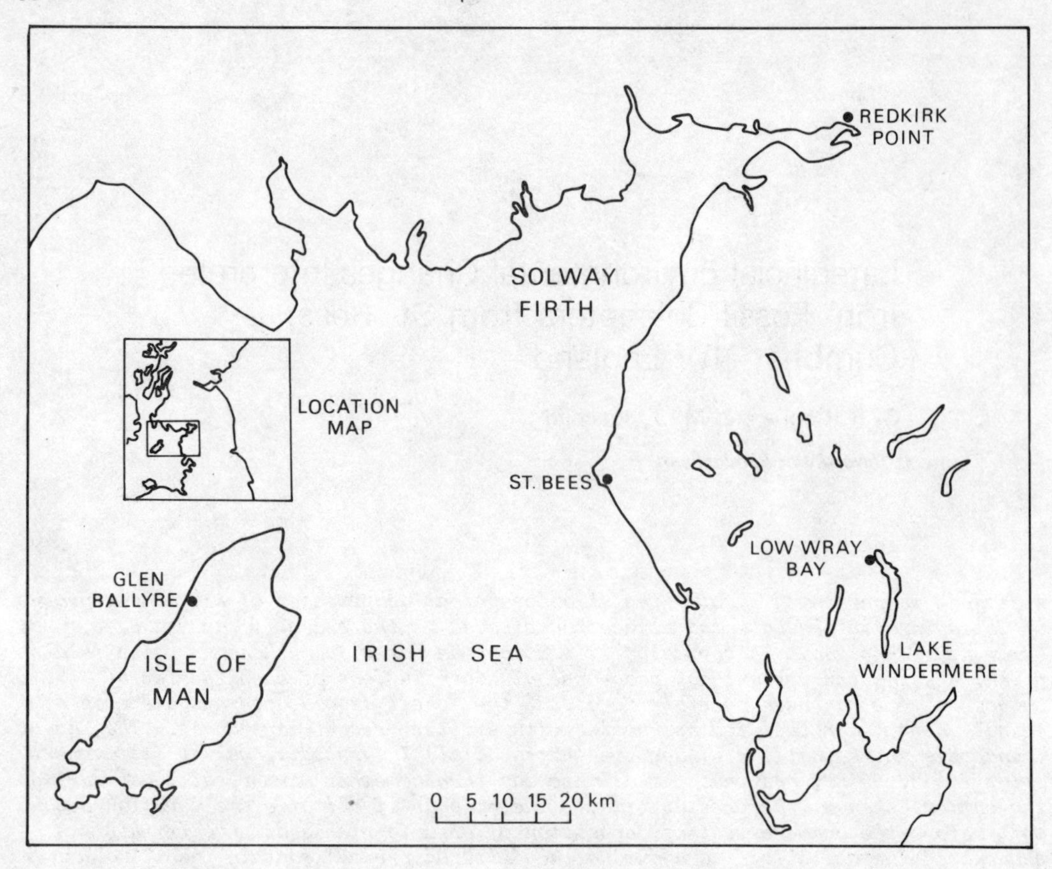

Fig. 1 Location of St. Bees and some other Lateglacial deposits

muds that once filled a hollow on the uneven surface of the drift to the SW of the town. These deposits are at present exposed about two-thirds of the way up the sea cliff, half a mile SW of St. Bees railway station (Nat. Grid Ref. NX965114). There can be no doubt that this is the same site that was investigated by Pearson, because he provides in his thesis (1960) a photograph with enough detail to allow precise identification of the locality. Furthermore, Pearson equates his locality with site 6 of Walker (1956). For the past decade this has been the only exposure of Lateglacial deposits in the neighbourhood, and during that time there has been little recession of the cliff. It seems likely that the section will continue to be available for reference, or for checking the results presented here, for at least another decade.

The stratigraphy of the Lateglacial deposits is straightforward. The most outstanding feature is the horizon of felted peat-like detritus mud which resists erosion and tends to stick out from the section which consists for the most part of sands with clay partings. The Lateglacial deposits dip inwards towards the centre of the basin; this is most obvious in the layer of detritus mud (Plate 1). This inward dip is either the result of subsidence of the hollow after the detritus mud was laid down, or a result of compaction of the highly organic sediment under the overburden of sands and clays. These overlying deposits also show evidence of post-depositional deformation in the form of festoons and folds indicating slumping

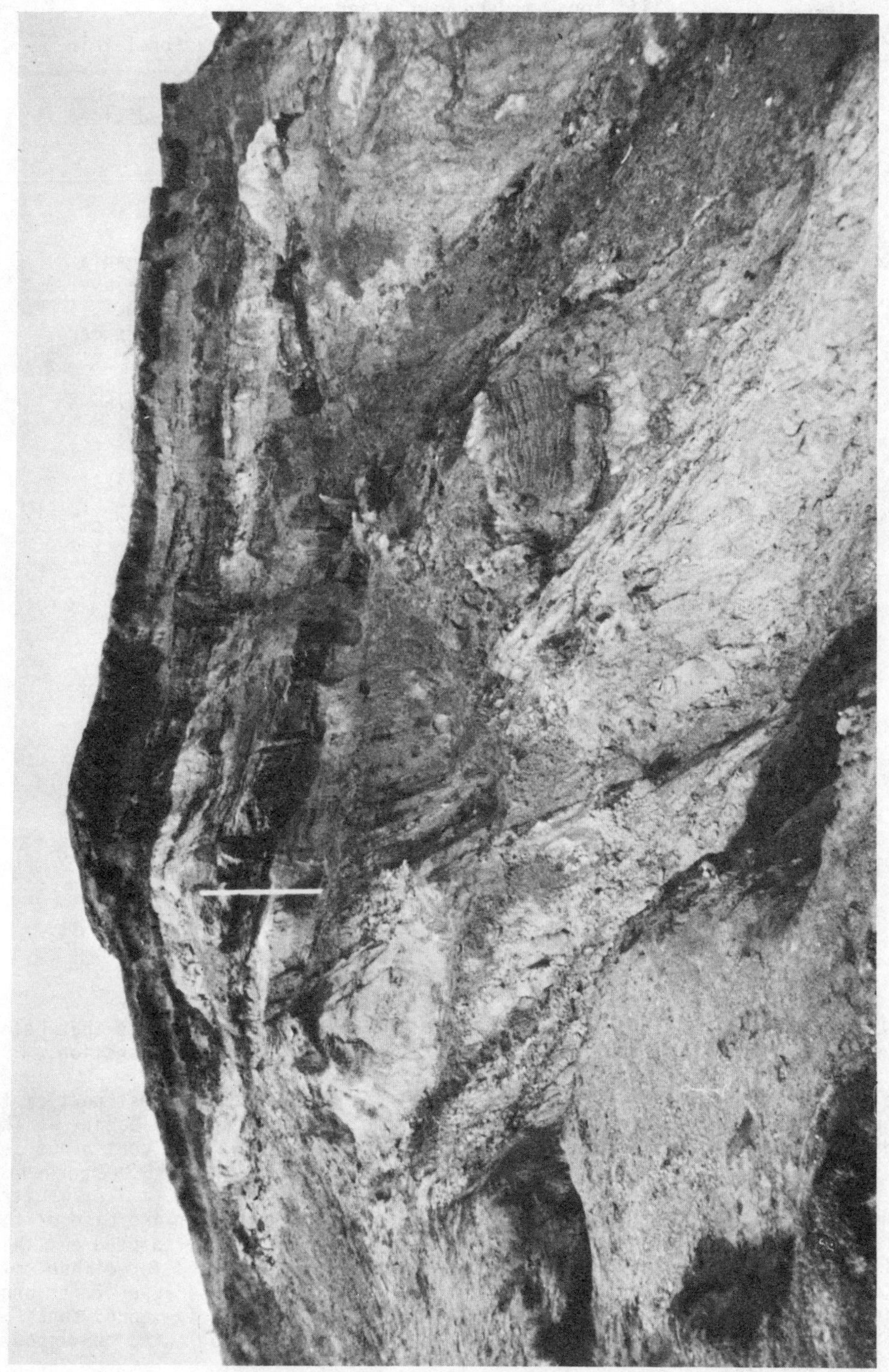

Plate 1 Cliff section in Lateglacial deposits at St. Bees, Cumbria.

416 cm	Cliff top; modern vegetation.
	Blown sand, crudely stratified, with occasional thin gravel layers. Slightly organic at base.
271 cm	
	Reddish-grey sand with thin layers of finely-divided organic material.
261 cm	
	Dark brown organic material with thin sandy intercalations.
254 cm	
	Sandy clay with considerable finely-divided organic content at top, grading down into paler, more clay-rich, less organic sediment at base. Occasional rounded pebbles. Vertical rootlets penetrate underlying material from this level.
232 cm	
	Crudely-bedded fine gravels, sands and silts with occasional coal fragments. Some cross-bedding.
142 cm	
	Alternating, sub-horizontal beds of sand and clay; some evidence of cross-bedding and channelling. Isolated rounded stones and coal pebbles up to 2 cm in diameter.
55 cm	
	Thin bed of grey clay.
50 cm	
	Dark brown detritus mud with abundant *Menyanthes* seeds at top and recognisable sticks at base. Impersistent thin sandy laminae throughout.
0 cm	
	Grey, silty, slightly organic sand, becoming pinker and less organic at depth. Occasional contorted organic laminae.
-40 cm	Grades into uniform red sand below -40 cm.

TABLE 1 SECTION THROUGH THE ORGANIC DEPOSITS AT ST. BEES

Datum level 0 cm is approximately 20 m above sea-level

towards the centre of the basin. Palaeontological evidence suggests that both subsidence and compaction played a role in the stratigraphy of the section.

The sampling site from which samples were obtained was in the deepest part of the exposed sequence, probably representing the middle of the original basin. The sedimentary sequence at the sampling site is given in Table 1. A continuous series of samples was taken, each of 5 cm in thickness and approximately 5,000 cc in volume, the only exception being the lowest sample which was 10 cm thick. An arbitrary sampling datum level "0" was taken at the sharp junction between the base of the detritus mud and the underlying grey silty sand. It should be pointed out that this basal sequence of arenaceous sediments was not investigated for either pollen or insect fossils by previous workers on this site, probably because of its unpromising appearance in the field. However, except for the crudely-bedded sands and gravels in the upper part of the sequence, all the samples collected contained insect fossils.

CHRONOSTRATIGRAPHY

A series of eleven samples for radiocarbon dating was taken from the same section that yielded the insect fauna. The oldest date was obtained from a bulk sample of the basal 5 cm of the detritus mud, and it apparently indicates that this deposit began to accumulate at 13,290 ± 310 BP (Birm-640). This date is in broad agreement with the commencement of organic sedimentation on the floors of the Cumbrian Lakes (Pennington, 1977a). However, a second date from exactly the same horizon and locality, but this time based on small broken sticks that were clearly laid down within the sediment and could not have penetrated from above, gave a radiocarbon age of 12,560 ± 170 BP (Birm-378). Furthermore, when a graph was plotted of radiocarbon age against the height of each sample above the base of the detritus mud (Fig. 2) (Williams and Johnson, 1976), two distinct groups of dates were displayed that form two linear series; the one made up of bulk samples of bottom sediment, the other comprising samples of terrestrial debris. Apparently the former yield radiocarbon ages that are about 500 years older than the latter. Since the increase in the apparent ages of subaquatic plant debris may be due either to dilution effects resulting from assimilation of dissolved carbonate derived from ancient limestone, or to the fractionation against the heavier isotopes of carbon by the

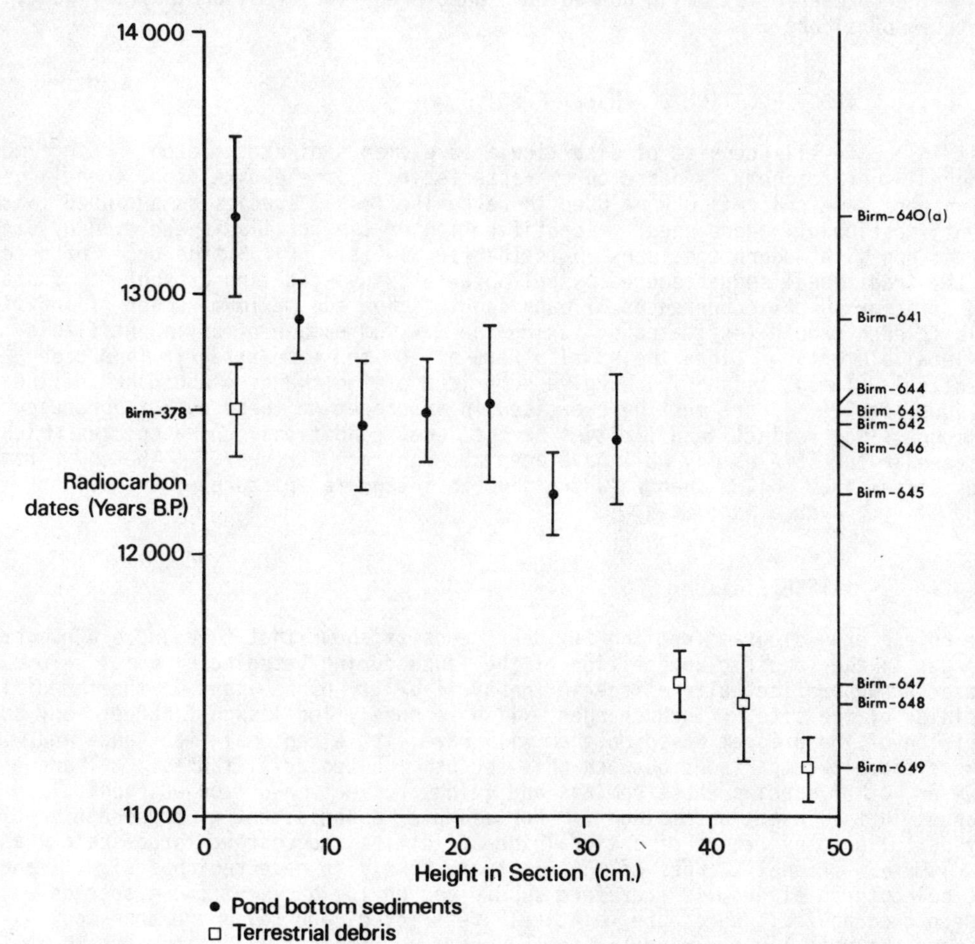

Fig. 2 Radiocarbon dates from the St. Bees deposits.

metabolic processes of the plants, the C13 values will not provide a measure of the discrepancy between the true age and the apparent age. Bearing this in mind, the only dates accepted here are those derived from terrestrial materials, namely Birm-378, Birm-647, Birm-648 and Birm-649 (see Fig. 2).

EXTRACTION OF INSECT FOSSILS

The fossils were recovered from the sediment by the standard technique developed in the Quaternary Laboratory, Dept. of Geological Sciences, University of Birmingham (Coope and Osborne, 1967; Coope, 1968). Briefly, the procedure is as follows. The sample is disaggregated and wet-sieved to remove any fine particles. The residue held on the sieves is then thoroughly mixed with paraffin in a polythene bowl. Water is introduced into the bowl in sufficient quantities to separate a floating fraction from the denser material on the bottom of the bowl. Because insect cuticle preferentially adsorbs the oil, the insect fossils tend to concentrate in the flotant while the bulk of the plant debris sinks. The floating fraction is then decanted and washed, first in a detergent solution and then in alcohol, after which the insect fossils are removed and sorted under a binocular microscope. The fossils are either preserved by being gummed down onto prepared cards, or they may be kept in tubes of alcohol.

IDENTIFICATION OF INSECT FOSSILS

Most insect fossils consist of disarticulated elements of exoskeleton. Since most modern insect taxonomy is based on characteristics of the exoskeleton, to a large extent the same criteria may be used to recognise fossil species as are used in the identification of modern ones. Identification of the fossils is achieved by direct comparison with modern specimens. In the faunal list (Fig. 3)* the order of species is the traditional sequence used by coleopterists investigating present-day faunas. The stratigraphical occurrences of each species give the maximum number of individuals in each sample (estimated by taking the maximum number of any identifiable skeletal element). Since the samples were all of the same size (in this case identical volume), these figures give some idea of the changing abundance of the various species. Care must be exercised in interpreting these numbers because commonness may reflect peculiarities of the local conditions, and a species which is rare in our samples may well have been abundant not far away. These diagrams thus differ from pollen spectra which give an integrated picture of the pollen productivity over a wide area.

INTERPRETATION

The coleopteran diagram from the St. Bees deposits shows that there were a number of changes in the specific composition of the fauna during Lateglacial time. Some of these changes reflect alterations in the availability of habitats in the immediate vicinity of the site. Such changes will have purely local significance; any subdivision of the diagram based on them will have little regional importance and will make difficult comparisons between this and other Lateglacial faunas. There are, however, other species whose comings and goings reflect major zoogeographical changes in the insect assemblages. For example, associations of temperate species may be followed by others of artic/alpine affinities, suggesting large-scale changes of climate. Faunal changes of this sort are likely to have regional significance. The coleopteran diagram is therefore subdivided on the basis of those species whose modern geographical ranges are relatively restricted, and which are thus most likely to have the most limited climatic tolerances. Greatest attention is paid to the presence or absence of these species rather than to changes in their abundance; this is because the numerical representation of a species is chiefly a measure of the local suitability of habitats. It is, nevertheless, encouraging when the numbers of individuals of a critical species are high because this is an indication that the species was thoroughly established in the neighbourhood and not an accidental vagrant that had come in from further afield (see, for example, the abundance of *Bembidion octomaculatum*).

The St. Bees coleopteran diagram is divided into three Faunal Units, each defined on the specific composition of the fossil beetle assemblage. These Faunal Units will be designated StB II, StB III and StB IV, beginning with the lowest. We have deliberately omitted to define a Faunal Unit StB I on the basis of the available fossils from this section, since we prefer to reserve this designation for a basal arctic/alpine assemblage of Coleoptera recognized in other Lateglacial sequences but as yet not present in this section. This procedure permits comparisons to be made with faunal units in other, more complete, Lateglacial sections.

Faunal Unit StB II

This faunal unit includes assemblages of Coleoptera from the lowest horizon sampled, namely -40 cm to -30 cm, up to and including the sample from +15 cm to 20 cm. The assemblages are made up entirely of temperate species, and in spite of the fact that these sediments represent the beginning of Lateglacial sedimentation at this locality, there are no arctic/alpine species present. The key species that characterise this faunal unit will be listed below when dealing with its climatic significance.

The local environment indicated by this faunal unit may be divided into two subunits. The basal sand contains a number of open-ground beetles, suggesting that the vegetation was sparse. There are conspicuously few dytiscid water beetles, which suggests that at this level, the sedimentary basin did not contain permanent water during the summer. With the increase in organic content of the sediment there was a sudden rise in the number of water beetles, suggesting that by this time a more or less permanent pond became established at the site. Surrounding this pond were abundant *Carex* spp., and further afield, weeds of open ground. The general impression given by the coleopteran assemblage is of a thin soil supporting a scant vegetation rather similar to newly-colonised waste tips round gravel pits. In many respects the insect fauna suggests a pioneer community rather than one which has reached equilibrium with its physical environment.

The species whose presence uniquely characterise this faunal unit are as follows: *Asaphidion cyanicorne, Bembidion assimile, B. octomaculatum, B. varium, Harpalus rubripes, Acupalpus dorsalis, Amara equestris, Cymindis angularis, Brachypterolus pulicarius, Airaphilus elongatus, Donacia marginata, Cleonus piger, Bagous limosus, Smicronyx coecus, Baris peramoena* and *Gymnetron beccabungae*. All these species have present-day distributions that are largely central and southern European. This faunal unit indicates that the climate of the time was, during the summer, at least as warm as, or maybe even warmer than it is in northern England today. This would mean that July average temperatures were probably near to 16°C.

Whereas almost all the constituent species of this faunal unit live today in temperate western Europe, one, *Baris peramoena*, is a steppe species from southern Russia. This single species hardly merits the designation of a steppe element in the fauna (the rest of the characteristic steppe assemblage is absent), and we believe that its presence in northern England at this time may be explained by the combination of warm summers and an open, largely treeless landscape, where ground conditions may have mimicked some of the steppe environments without any implication of a steppe-type climate.

Faunal Unit StB III

The base of this faunal unit may be taken at +20 cm where the last of the 'southern' elements listed above die out. They are replaced by a gradually increasing number of species whose present-day geographical ranges are predominantly northern. The uppermost boundary of this faunal unit is not so easy to define because there is a gradual increase in the number of boreal species. An arbitrary boundary at +45 cm has therefore been selected.

There is abundant evidence from the Coleoptera that the deposit at this level represents the infilling of a eutrophic pond. The Dytiscidae suggest that open water was available, and the abundant Gyrinidae (the familiar whirligig beetles) indicate that the surface was not covered by floating vegetation. The pool supported a submerged plant community, the habitat for the subaquatic weevils *Eubrychius velatus* and *Litodactylus leucogaster*. Emergent vegetation in the form of reeds and sedges must have been plentiful to support the abundant representation of *Donacia* spp. Woodland species of Coleoptera are present in this faunal unit, but it is difficult to determine what species of tree are involved. *Salix*, at least, is indicated by the presence of *Melasoma collaris*, *Cryptocephalus punctiger*, *Caenorhinus tomentosus*, *Rhynchaenus* spp. and *Rhamphus pulicarius*.

The climatic implications of this faunal unit contrast with those of Faunal Unit StB II. The total disappearance of the southern species from the coleopteran diagram cannot be explained by invoking some change in the local environment as the species involved are too diverse in their ecological requirements. Furthermore, there is a complementary incoming of species whose geographical ranges are chiefly northern, and again these species collectively span a wide range of environmental preferences. They include *Patrobus septentrionis*, *Agonum consimile*, *Agabus arcticus*, *Arpedium brachypterum*, *Otiorhynchus nodosus* and *Notaris aethiops*. There can be little doubt that this faunal change is indicative of a marked climatic deterioration. Average July temperatures must have been at about 13°C. Such a thermal regime would be quite compatible with the existence of *Betula* woodland suggested by the palynological record for this episode (Walker, 1956). Indeed, the birch woodland could have continued to expand in spite of the deterioration in climate, since the temperatures involved do not yet approach the lower limit for its survival.

Faunal Unit StB IV

The base of this unit is taken at +45 cm where the number of boreal species suddenly increases and the number of individuals of this group like-wise rises. The upper limit is set by the incoming of largely unfossiliferous stony sands and clays, capped by recent blown sands. A single horizon of grey organic clay at +232 cm was found to contain fossil insects entirely in keeping with this faunal unit. No deposits transitional to the Flandrian were found in this section when it was sampled.

The Coleoptera show that the eutrophic pond must have been filled in by sediment near the beginning of this faunal unit, and replaced by cold, boggy pools, the habitat of *Deronectes griseostriatus*, *Agabus arcticus*, *A. congener* and *Dytiscus lapponicus*. The carabid species *Patrobus assimilis* and *Pterostichus diligens*, and the polyphagous weevil *Otiorhynchus nodosus*, are all present in sufficient numbers to suggest that acid environments were locally available, but it is interesting to note that at horizons above +100 cm the numbers of these species decline. Perhaps this is an indication that, by then, the frost disturbance of the soil was changing its base status. In the damper places, there was still abundant *Carex* (indicated by *Notaris aethiops*), and *Salix* bushes (indicated by *Phyllodecta vitellinae*). The weevil *Deporaus betulae*, in spite of its name, attacks leaves not only of birch but also of willow. Many of the small omaliine staphylinids (*e.g. Olophrum fuscum*,

Arpedium brachypterum, Boreaphilus henningianus), so common particularly at the base of this unit, are predators on other soil arthropods under leaf litter and moss. The abundance of *Simplocaria metallica* also indicates that there was at this time plentiful moss, upon which both adult and larva feed. Finally, in the drier places, the vegetation was sparse and patchy, as indicated by the ground beetles *Amara quenseli* and *A. alpina.*

Climatically this faunal unit shows evidence of still further deterioration, with an increase in the numbers of northern species over those already represented in the previous faunal unit, and the incoming of a marked arctic/alpine element. These species include *Diacheila arctica, Bembidion mckinleyi, Amara alpina, A. quenseli, A. torrida, Deronectes griseostriatus, Hydroporus arcticus, H. lapponum, H. tartaricus, Colymbetes dolabratus, Helophorus sibiricus, Olophrum boreale, Pycnoglypta lurida, Boreaphilus henningianus, Simplocaria metallica* and *Otiorhynchus nodosus.* Such a faunal assemblage is indicative of harsh climatic conditions, with average July temperatures near to 10°C. The abundance of the fauna, however, suggests that temperatures were probably not much below this level. A thermal regime such as this would lead to widespread elimination of the birch woodland, but it would allow the survival of tree birches in favoured localities whose aspect caught the sun and gave shelter from the wind.

CLIMATIC INFERENCES: SUMMARY AND DISCUSSION

The interpretation of Quaternary palaeoclimates from biological data relies almost entirely on our understanding of the present-day geographical ranges of the species concerned. This applies whether we are dealing with plants or animals, and whether they are terrestrial or aquatic, marine or fresh-water organisms. Since climate is described in regional terms and its parameters measured in averages, it is the broad range of a species that provides information on its climatic preferences. It is the macroclimate that provides the arena in which each species acts out its life in the intricate world of microclimate. The fact that a particular species of beetle has exceedingly precise temperature and humidity requirements that dictate that it can live under this stone but not that, or on this side of a lettuce leaf rather than on the other, is absolutely vital to the insect but it tells us little about the climate. Likewise, precise laboratory experiments on the environmental tolerances of organisms or sophisticated field studies on the microhabitats of species will tell us much about the biology of plants and animals, but provide us with little useful information from which to interpret palaeoclimates.

It is for the above reasons that we look to the broad geographical distributions of our species of Coleoptera for clues to past climates. There are good entomological reasons for believing that the limits of these ranges are set by factors in the physical environment (Price, 1975; Thiele, 1977; Coope, 1977), whereas the abundance of a species within its range is determined more by biotic factors that are density-dependent. In our interpretation of the palaeoclimatic significance of a species of beetle, greatest attention is therefore paid to the limits of its present-day range, and the broad environmental conditions along those limits. In our interpretations of coleopteran spectra we pay particular attention to the presence or absence of a species rather than to its abundance.

To reduce to some sort of order the rather oppressive wealth of beetle data from our fossil localities it is necessary to group together biogeographically similar species. In the case of the St. Bees fauna, seven distributional categories have been established. It must be emphasised that membership of any category was determined entirely on the basis of the known present-day geographical distribution of a species, and that no notice was taken of its fossil occurrences. The distributional categories are as follows:-

A: Boreal and montane species whose normal range is above the tree-line.

B: Boreal and montane species whose normal range also includes the upper part of the coniferous forest.

C: Widespread species whose normal range is north of central Britain.

D: Cosmopolitan species.

E: Widespread species whose normal range is south of central Britain.

F: Southern species whose northern limit of distribution just reaches, or just fails to reach, southern England.

G: Southern European species.

Figure 4 shows distribution maps for a representative species of each category, with the exception of category D.

Figure 5 shows the numbers of species in each of these categories at all sampled levels of the St. Bees deposit. Category D has been omitted from this diagram because these very widespread species are seemingly tolerant of a considerable range of climates. From the point of view of palaeoclimatic studies, category D may be relegated to the role of 'background noise'. The actual species included in each category (except D) are shown by the appropriate letter at the foot of each species column in the coleopteran diagram (Fig. 3); unlabelled species are in category D.

It is clear from Fig. 5 that species belonging to the more southern categories are confined to the basal part of the sequence, and that at these levels the northern species are almost totally absent. The presence of category C species in the basal parts of the detritus mud is of particular interest, suggesting that something of a climatic deterioration had set in during the deposition of samples +10 cm to +20 cm. It should be borne in mind, however, that many category C species live today in Britain as far south as St. Bees, and that there need have been little climatic change to permit their arrival. In all other respects the assemblage of Coleoptera from Faunal Unit StB II is typical of temperate conditions. There is no way of dating precisely the beginning of this temperate episode at St. Bees, apart from pointing out that southern category species occur 40 cm below the base of the detritus mud that dates at 12,560 ± 170 BP. Bearing in mind that the rate of sedimentation of the grey sand was no doubt faster than that of the detritus mud, it seems likely that temperate climates already existed at St. Bees by 13,000 BP. It is therefore interesting to note that the surface waters of the North Atlantic Ocean off our west coast had already by this time achieved temperatures equivalent to those of the present day, according to Ruddiman *et al.* (1977). Since the ocean currents are largely wind-driven, the establishment of an ocean circulation in our area, so similar to that now in existence, implies an atmospheric circulation that likewise resembled that of the present day. It is difficult to escape the conclusion that at this time the climate of northern England was as temperate as it is today. Because of the open, rather treeless landscape, some sunny places might well have experienced temperatures high enough to permit some of the category F and G species to flourish well to the north of their present-day geographical limits.

In the light of the climatic inferences discussed above, the largely treeless environment might at first sight seem anomalous. The herbaceous flora was however diverse at this time (Birks and Deacon, 1973), covering a wide range of phytogeographic groupings. This is clearly no picture of a tundra vegetation. Furthermore the absence of trees cannot be blamed entirely on migration rates that were inadequate to keep pace with the rapidity of the climatic amelioration as suggested by Coope (1970b), because macroscopic remains of *Betula pubescens* were recorded from

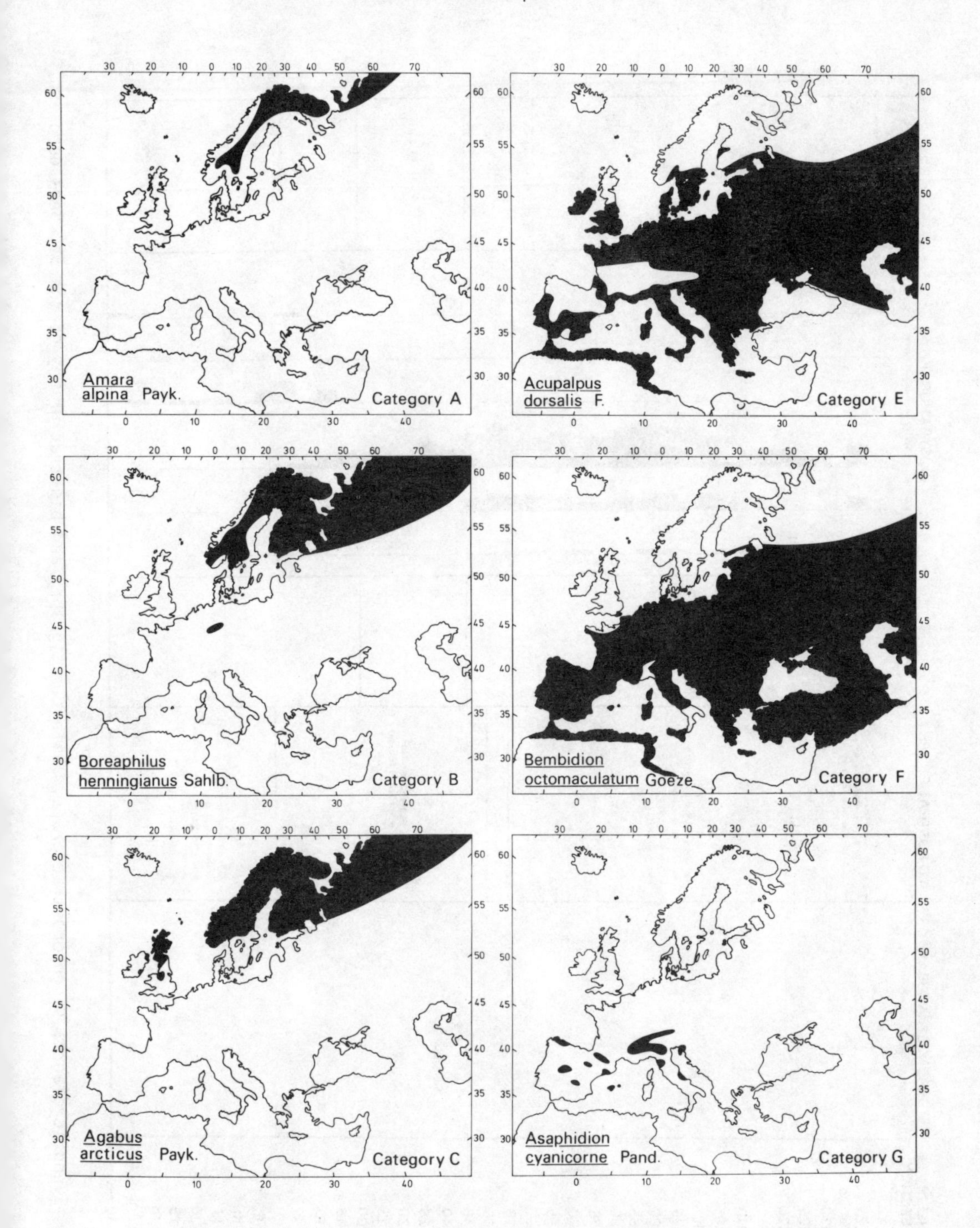

Fig. 4 Distribution maps for representative species in each biogeographical category (Category D excluded).

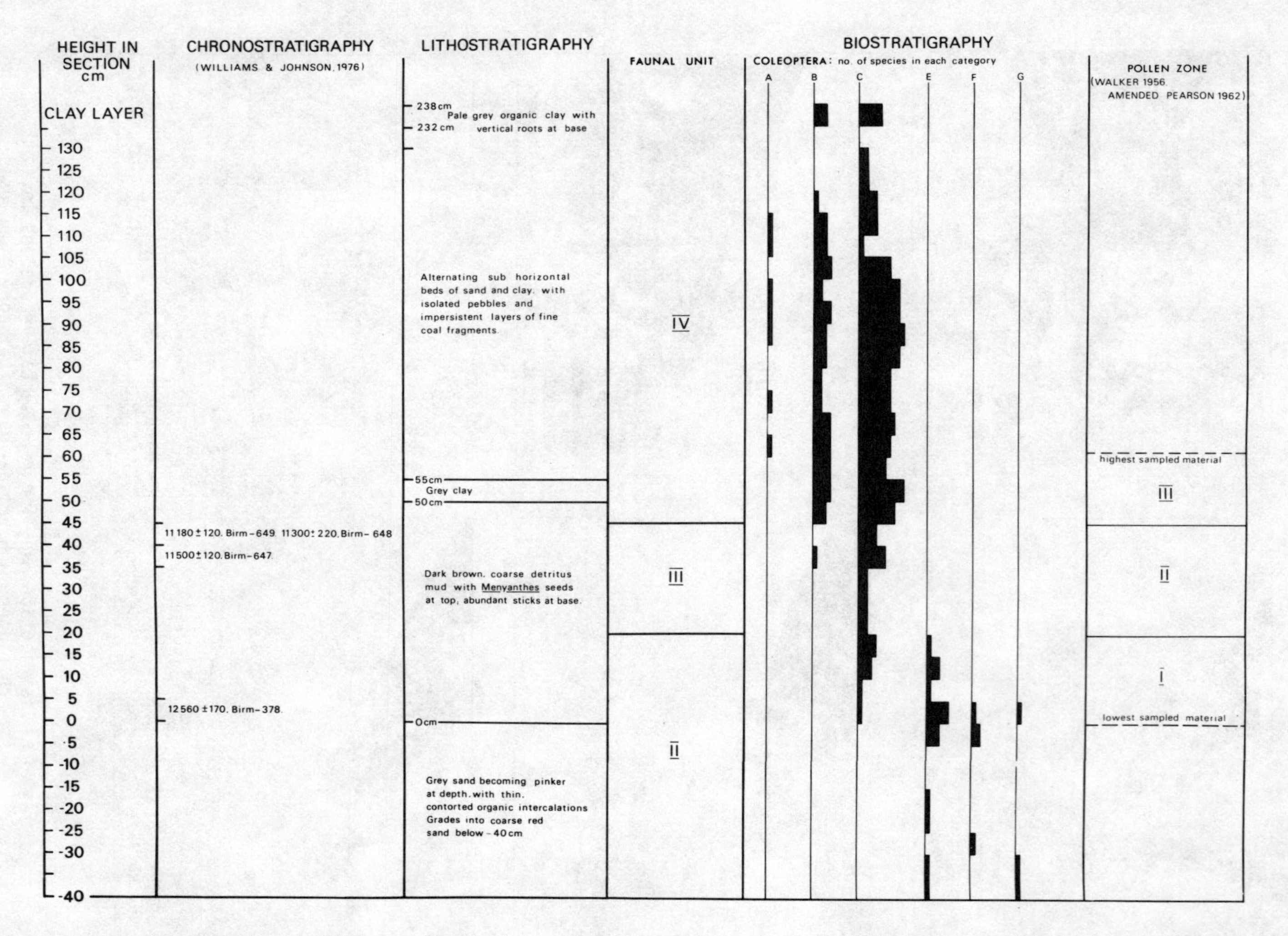

Fig. 5 Synoptic diagram of chronostratigraphy, lithostratigraphy and biostratigraphy of the St. Bees deposits.

deposits of equivalent age at Low Wray Bay, Windermere (Pennington, 1977a). What, therefore, were the factors that conspired to prevent birch woodland becoming established in a climatic regime that was entirely temperate?

We offer two contributory factors whose combined effect might well have been to inhibit the development of birch woodland in the early Lateglacial period. The first of these is a commonly suggested factor that will be dealt with here only briefly. It is the old problem of the state of the soil. Since almost all the northern parts of the British Isles had been but recently abandoned by the ice sheets, soil development had to start from scratch with raw rock debris. Even south of the Devensian ice-limit, frost heaving during the maximum glaciation was intense (Williams, 1975), and this would have destroyed much or all of the soil structure. The sparse vegetation cover would have meant that away from the poorly drained valley bottoms there was little recruitment of organic matter to these incipient soils and any precipitation would flow freely through them. Certainly the vegetation cover and insect assemblages of this period resembled the present-day communities that colonise areas in England from which the topsoil has been stripped. However, these are just the places where *Betula* scrub develops today and it thus seems that raw freely-draining soil is insufficient by itself to inhibit the growth of birch woodland. Some other factor must also be involved.

The second contributory factor is the climate. The insect fauna of this period strongly suggests that summers were as warm as, or even warmer than, those of central Britain at the present day. The warm summer of 1976 shows how vulnerable even mature birch trees are on well drained soil, particularly in the southern parts of this country. This particular summer also encouraged many insect species to explore new extensions of their ranges well to the north of their accustomed limits. Summers of lesser intensity would have had equally drastic effects on seedling regeneration. It is now evident from analysis of beetle assemblages of the early part of the Windermere (Lateglacial) Interstadial that summers of this sort were to be expected, not necessarily on an annual basis but with sufficient frequency to tempt the most southern species of Coleoptera (categories F and G) to extend their ranges at least as far north as St. Bees where they are found in company with an otherwise thoroughly English assemblage of species. Warm summers coupled with freely draining soils of low organic content may well have inhibited the spread of birch woodland at this time leading to the establishment of a varied flora and fauna that has no exact analogue at the present day; not because the climatic regime has no equivalent but because accidents of space and time dictated that combinations of habitats were available then that are not available at the present time. These peculiar environmental conditions are not to be looked on as 'extinct habitats', since they are likely to recur whenever similar circumstances arise.

If this speculation is correct it casts new light on another anomaly of Lateglacial palaeoenvironmental interpretation, namely the apparent earliness of the response of the flora in the NW of Britain rather than in the SE of the country where it might have been expected. The important observation by Pennington (1977a, p.268) that "In general, the evidence from (north) western sites suggests an interstadial vegetation type containing varying proportions of *Betula, Juniperus* and *Empetrum,* which both developed earlier and represented a different vegetation type, when compared with the birch-pine p.a.z. of eastern England, which reached its maximum development in Alleröd time" is difficult to reconcile with the traditional view that the rise of *Betula* in Lateglacial pollen diagrams is an indication of climatic warming. If the effect of pre-Alleröd high summer temperatures was to inhibit the expansion of *Betula* woodland this effect would have been greatest in the south and east and the early expansion of birch woodland might have been expected in the north and west of the British Isles, provided that the tree birches could have gained access to the north from their glacial refuges, presumably well to the south of the British Isles. Their route northwards may well have lain along the valley bottoms that were immune from the dessicating heat of the summers.

The coincidence of timing between the rise of *Betula* pollen and the evidence of cooler summers inferred at St. Bees has been found at a number of Lateglacial sites: Bodmin Moor (Brown, 1977; Coope, 1977), Glanllynnau (Simpkins, 1974; Coope and Brophy, 1972), Church Stretton (Rowlands and Shotton, 1971; Osborne, 1972), and at Redkirk Point (Bishop and Coope, 1977). At Glen Ballyre, on the Isle of Man, there is a similar slight rise in *Betula* (Dickson *et al.*, 1970; Joachim, 1978). The wide spread of latitude between all these localities suggests that there is some causal connection between the climatic cooling and the spread of birch woodland.

It is always pleasant to finish on a note of total agreement. The climatic implications of Faunal Unit StB IV are in complete accord with the botanical, sedimentological and oceanic evidence, all pointing to a climate of glacial severity. In spite of the shortage of time, the great thickness of these deposits testifies to the vigour of the geomorphological activities that filled the sedimentary basin and brought the Lateglacial record at St. Bees to an end.

* NOTE

Figure 3 is included inside the back cover of the book.

A Lateglacial Site at Stormont Loch, Near Blairgowrie, Eastern Scotland

C. J. Caseldine

(University of Exeter)

ABSTRACT

The analysis of two adjacent profiles from Stormont Loch near Blairgowrie on the southern fringe of the eastern Grampian Highlands demonstrates the existence of a comparable sequence of local pollen assemblage zones covering the Lateglacial period. Relative pollen counts are supplemented by the determination of pollen concentration values expressed as the number of grains per unit volume of sediment for one of the profiles. The sequence of local pollen assemblage zones is dissimilar to that found at most other sites in the Grampians for there are two zones dominated by woody species, Juniperus - Empetrum and Betula - Juniperus respectively, which are separated by a Rumex - Cyperaceae zone. Total pollen concentration varies from over 30×10^3 grains per cm^3 in the zones dominated by woody species to only 10×10^3 grains per cm^3 in the Rumex - Cyperaceae zone implying either an increased sedimentation rate or reduced pollen productivity at this time. The occurrence of such a sequence is not considered to be due to any stratigraphical discontinuities as there is no change in sediment lithology, but is thought to represent a significant change in the local vegetation communities with the disappearance of Juniperus and Empetrum. It is suggested that the sequence is comparable to that found by Walker at Corrydon in Glenshee (1977) and may represent the Bölling - Older Dryas - Alleröd chronozones. The vegetation record of the later Loch Lomond Stadial is characterised by three local pollen assemblage zones with high values for Salix, Rumex, Thalictrum and Artemisia, the latter becoming more important towards the later stages of the stadial due perhaps to a combination of increased dryness and the loss of the interstadial soil cover.

INTRODUCTION

Lateglacial profiles along the southern and eastern margins of the Grampian Highlands have been analysed by several authors (Vasari and Vasari, 1968; Vasari, 1977; Walker, 1975a, 1977; Lowe and Walker, 1977) but as yet there are no published diagrams for lower Strathmore (Fig. 1). Investigations of deposits in Stormont Loch near Blairgowrie (ref. NO1942) indicated the presence of sediments of Lateglacial age and the results of the analysis of these sediments using both relative pollen counting and the determination of pollen concentration values are presented here. Stormont Loch lies at an altitude of *c*. 61 m in one of a series of kettleholes within fluvioglacial sands and gravels of Highland origin on the outer margins of the Blairgowrie outwash fan (Paterson, 1974). This fan was produced during the

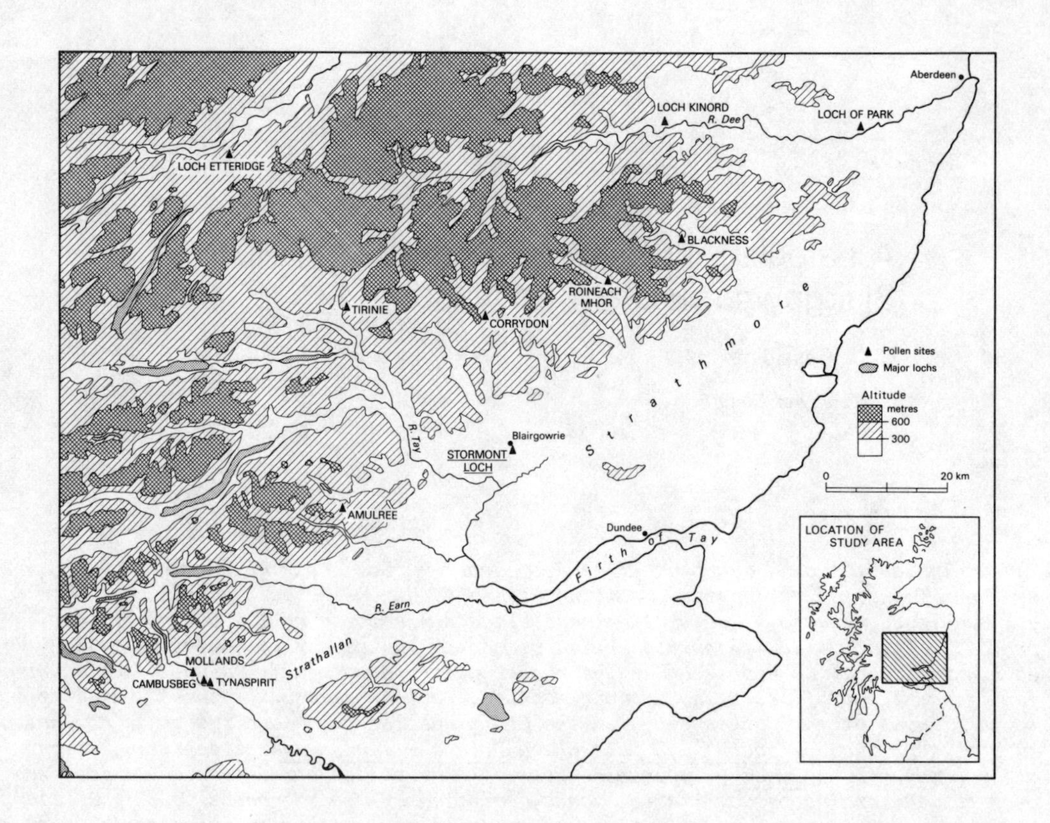

Fig. 1 Location map of sites in eastern Scotland

wastage of the main Devensian ice sheet by meltwater from Strathardle and Glenshee which entered Strathmore through the gorge in which the River Ericht now flows.

STRATIGRAPHY AND SAMPLING

Two cores were taken for analysis from the western edge of the loch within *Salix* carr which has developed over a subsidiary basin. The first core, SGI, was taken with a Russian borer at a depth of 466-516 cm and the second, SGII, was taken 10 m to the north of the first core at the deepest sampleable point reaching 620 cm. Although at different depths, the basic stratigraphy was similar with an organic mud sandwiched between silty clays and clays. The stratigraphy at SGI was as follows:

469 - 482 cm	Coarse grey clay fining with depth and with few observable organic remains
482 - 485.5 cm	Transition between fine and very fine clay with increasing organic content
485.5 - 486.5 cm	Thin dark brown organic band

486.5 - 488 cm	Coarse grey sandy silt
488 - 488.5 cm	Fine grey clay
488.5 - 492 cm	Fine grey green silty clay
492 - 498 cm	Dark brown organic mud with a slightly micaceous appearance becoming almost peaty towards the base
498 - 500 cm	Light grey-green gyttja
500 - 511 cm	Organic mud with occasional fruit stones of *Potamogeton* and faint silty bands
511 - 514 cm	Light grey silty clay with organic flecks
514 - 516 cm	Dark grey clay becoming increasingly coarse and gravelly with depth.

At SGII the stratigraphy was as follows:

550 - 556 cm	Light brown sedge peat with a micaceous appearance at the lower transition and fruit stones of *Potamogeton*
556 - 557 cm	Grey silty clay
557 - 566.5 cm	Fine grey silty clay with organic flecks
566.5 - 591.5 cm	Very fine silty clay becoming less silty between 566.5 cm and 581.5 cm with few organic remains
591.5 - 592.5 cm	Coarse grey sandy silt
592.5 - 604.5 cm	Brown organic mud
604.5 - 605.5 cm	Thin peat band with no observable minerogenic content
605.5 - 617.5 cm	Grey silty clay
617.5 - 620 cm	Dark grey clay increasing in coarseness with depth.

Samples were taken from SGI at intervals of 1 cm or less depending on the stratigraphy but for SGII the sampling interval was extended to 2 cm. All samples were treated with 10% HF prior to the use of Erdtman's acetolysis and mounted in silicone oil, but for SGII known assays of exotic spores in tablet form were added to a known volume of sediment for the determination of pollen concentration values (Stockmarr, 1971). The pollen diagram for SGI is based on percentages of a Total Land Pollen (TLP) count of 350 grains (Figs. 2a,b,c), whereas the diagrams for SGII are expressed either as percentages of the TLP count (Figs. 3a,b) or as pollen concentration in the number of grains per cm^3 of sediment (Figs. 4a,b,c) using a count of 500 grains including both TLP and exotic spores.

LOCAL POLLEN ASSEMBLAGE ZONES

Local pollen assemblage zones are defined for each profile as follows:

(a) SGI (Figs. 2a,b,c)

SG-1 Gramineae - *Rumex* Zone (516 - 512.5 cm)

The upper boundary which is marked by increasing *Juniperus* and *Empetrum*, and decreasing Gramineae and *Rumex*, precedes the lithostratigraphical change from a silty clay to organic mud. Gramineae dominates at up to 50% with high frequencies for *Rumex*, mainly *Rumex acetosa* type pollen.

SG-2 *Juniperus* - *Empetrum* Zone (512.5 - 505.5 cm)

This zone spans the lower part of the predominantly organic deposits and shows successive peaks for *Empetrum, Juniperus* and *Betula* but Gramineae, Cyperaceae, *Artemisia, Rumex* and *Thalictrum* are still well represented. The upper boundary is marked by increasing *Rumex* as *Juniperus* and *Empetrum* decline.

SG-3 *Rumex* - Cyperaceae Zone (505.5 - 499.5 cm)

Values for *Juniperus* are reduced to less than 1% but there is a small peak for *Pinus*. At the end of the zone NAP comprises over 90% prior to a sharp reduction in values marking the upper boundary and consists largely of Gramineae, Cyperaceae and *Rumex*.

SG-4 *Betula* - *Juniperus* - *Filipendula* Zone (499.5 - 490.5 cm)

Both *Betula* and *Juniperus* reach their highest peaks as does *Empetrum*. Most of the herb species characteristic of earlier zones are reduced or absent but *Filipendula* is consistently present as is *Potamogeton*, the main element of the aquatic flora. The change to silty clay precedes the pollen zone boundary which is defined where there is a sharp decrease in the pollen of woody species.

SG-5 *Salix* - *Rumex* - *Selaginella* Zone (490.5 - 485.25 cm)

Values for *Juniperus* are severely reduced but *Betula* is consistently present at *c.*10%. *Rumex* returns to high levels and there are continuous curves for Caryophyllaceae and *Thalictrum* with increasing *Salix, Pinus* and *Artemisia*. The upper boundary is marked by a sharp increase in *Artemisia*.

SG-6 *Salix* - *Artemisia* - *Rumex* - *Thalictrum* Zone (485.25 - 470.5 cm)

There are high values for *Salix*, Cyperaceae, *Artemisia, Rumex* and Gramineae and there is an increase in the number of taxa represented with continuous curves for Caryophyllaceae, *Thalictrum* and Rosaceae.

SG-7 Gramineae - Cyperaceae - *Rumex* - *Thalictrum* Zone (470.5 - 468 cm)

The change to SG-7 is marked by reductions in *Salix* and *Artemisia* as Gramineae and Cyperaceae increase with a gradual improvement in pollen preservation.

Statistical zonation of this profile using methods devised by Gordon and Birks (1972b) confirmed the existence of the boundaries outlined above except for the division between SG-1 and SG-2 which was not found by any of the procedures used.

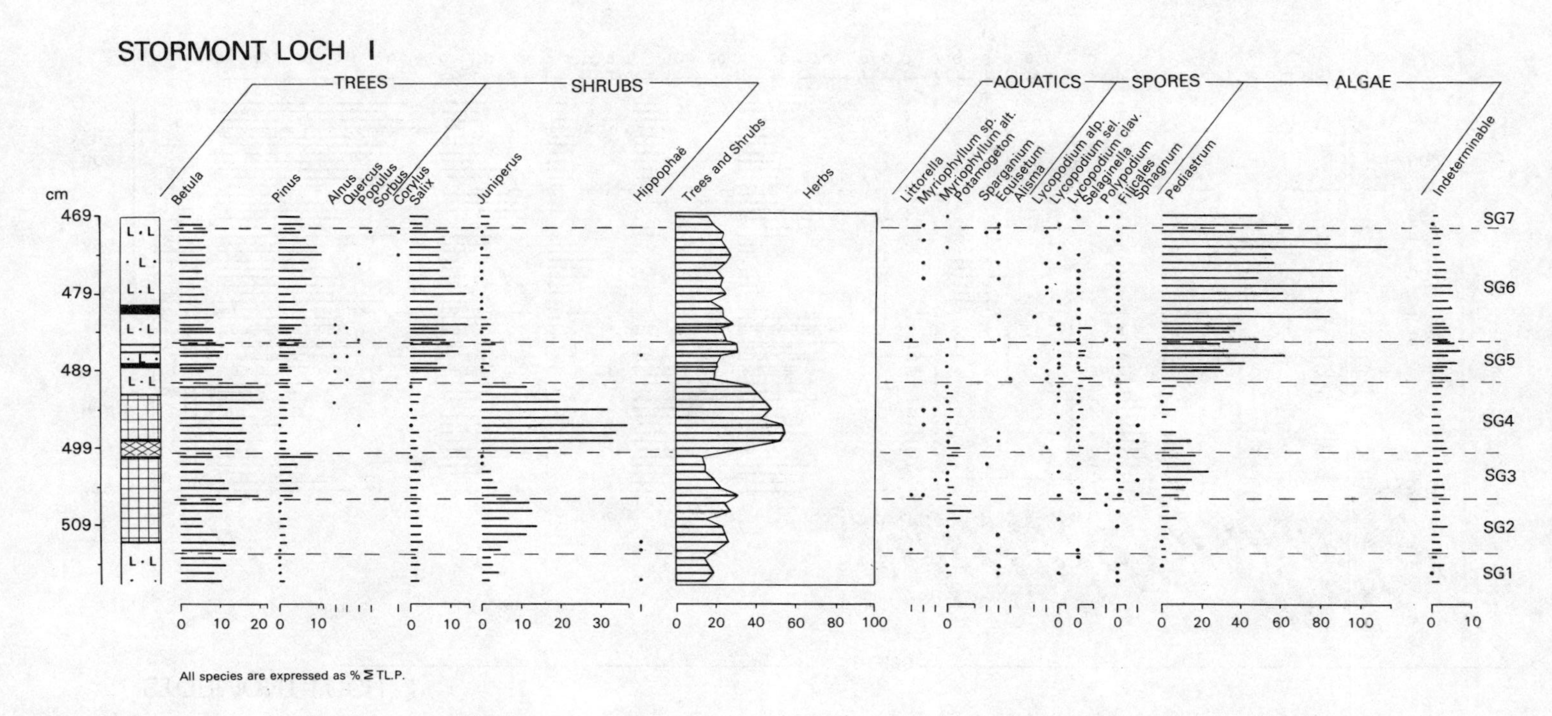

Fig. 2a Relative pollen diagram for SGI showing trees, shrubs,aquatics, spores and algae.

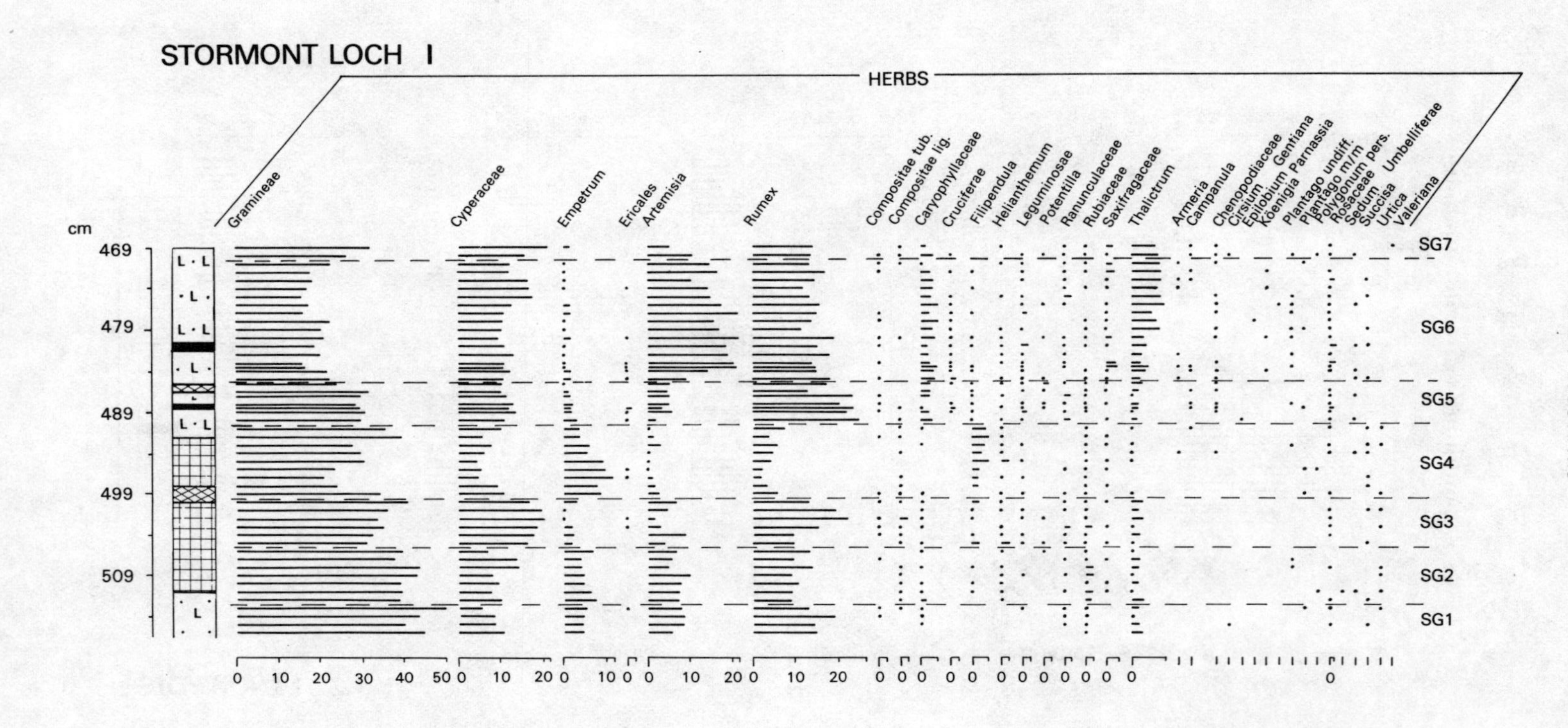

Fig. 2b Relative pollen diagram for SGI showing herbs only.

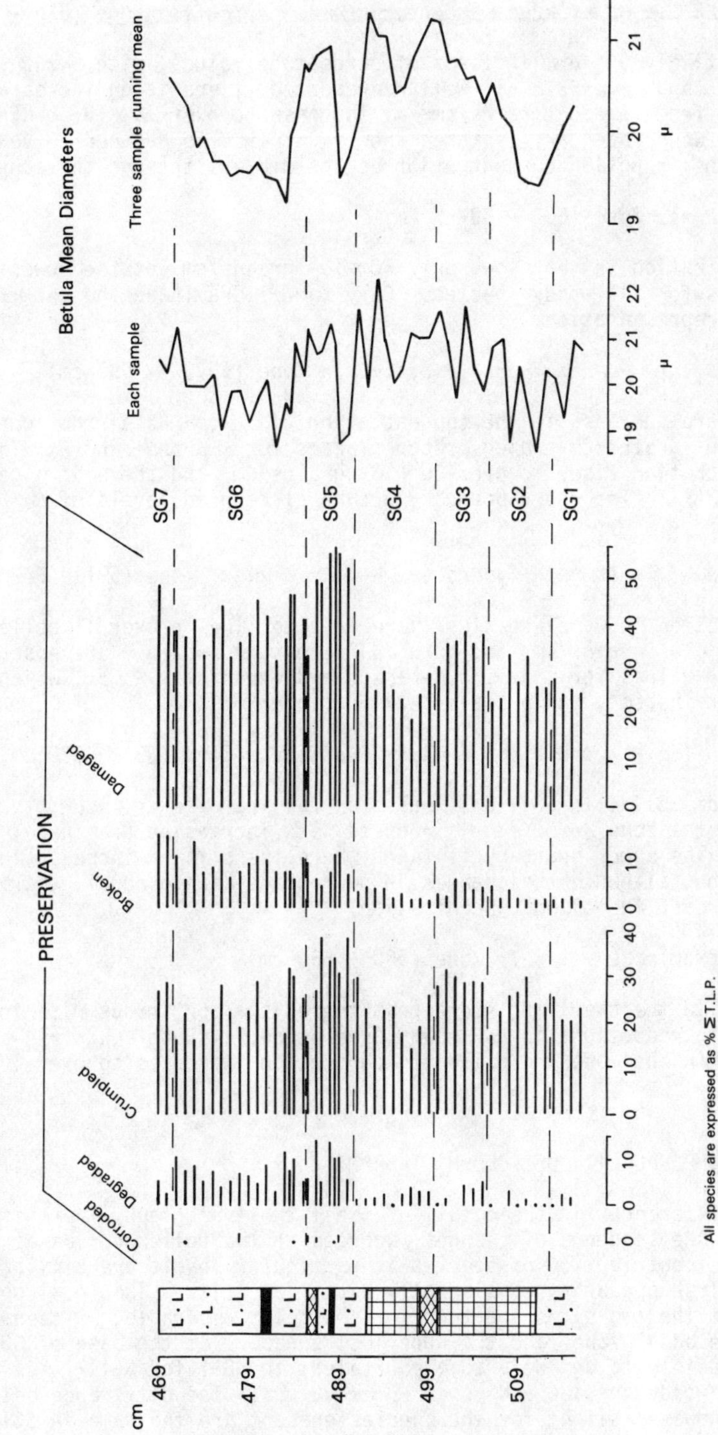

Fig. 2c Relative pollen diagram for SGI showing pollen preservation and *Betula* grain size.

(b) SGII (Figs. 3a,b and 4a,b,c)

SGII-1 *Betula* - *Juniperus* - *Empetrum* - *Rumex* - *Artemisia* Zone (616 - 602 cm)

Although relatively uniform in terms of percentage values, with *Betula, Juniperus, Empetrum, Rumex* and *Artemisia* all well represented, there is an increase in pollen concentration from 15×10^3 grains/cm^3 at the base to over 36×10^3 grains/cm^3 at 612 cm. The upper boundary is taken where *Juniperus* is reduced to very low values but the decline in pollen concentration begins mid-way through the zone.

SGII-2 *Rumex* Zone (602 - 597.5 cm)

Pollen concentration is very low, only 10×10^3 grains/cm^3 at the lowest levels with reduced values for all woody species. Only *Rumex*, Gramineae and *Artemisia* show any increases in representation.

SGII-3 *Betula* - *Juniperus* - *Filipendula* Zone (597.5 - 591 cm)

There is a sharp increase in the concentration of *Juniperus* at the lower boundary to over 11×10^3 grains/cm^3 which precedes peaks for *Empetrum* and *Betula*. Overall pollen concentration rises to over 30×10^3 grains/cm^3 and there is a general reduction in NAP. The upper boundary occurs at reduced levels of *Betula* and *Juniperus*.

SGII-4 *Salix* - *Rumex* - Cyperaceae - *Selaginella* Zone (591 - 582 cm)

With the reduction in woody species there is a decline in overall pollen concentration to 21×10^3 grains/cm^3 and both *Juniperus* and *Empetrum* are absent. There are increasing values for *Salix*, Cyperaceae and especially *Rumex*, which reaches over 9×10^3 grains/cm^3, and a small peak for *Selaginella*.

SGII-5 *Salix* - *Artemisia* - *Rumex* - *Thalictrum* Zone (582 - 560 cm)

The change from SGII-4 to SGII-5 is based on the increase in *Artemisia* but in the upper part of the zone *Artemisia* is reduced with increasing Gramineae, *Rumex* and Cyperaceae. The upper boundary is taken where there is an increase in *Betula*. Within the zone, although the changes in percentages are gradual, there is almost a twofold increase in pollen concentration.

SGII-6 Gramineae - *Betula* Zone (560 - 550 cm)

At the change to the overlying sedge peat there is a continuous rise in Gramineae and *Betula* with reductions in *Rumex* and Cyperaceae. Both *Juniperus* and *Empetrum* also rise within the zone and pollen concentration increases to over 40×10^3 grains/cm^3.

COMPARABILITY OF POLLEN DIAGRAMS

Despite the differences in the detail of the lithostratigraphy there is very little difference in the sequence of changes recorded in the pollen curves at the two sites. The percentage contribution of species at particular levels are similar, only Cyperaceae values appearing consistently higher at SGII. The local pollen assemblage zones at the two sites are therefore largely comparable, as seen in Table 1, except for the basal zones and the uppermost zones. At the base of SGII it was not considered possible to define a zone equivalent to SG-1 for SGII-1 has a higher proportion of woody species and lower *Rumex* values. The difference between the two sequences is however slight for the species present are the same in SGII-1 as SG-1

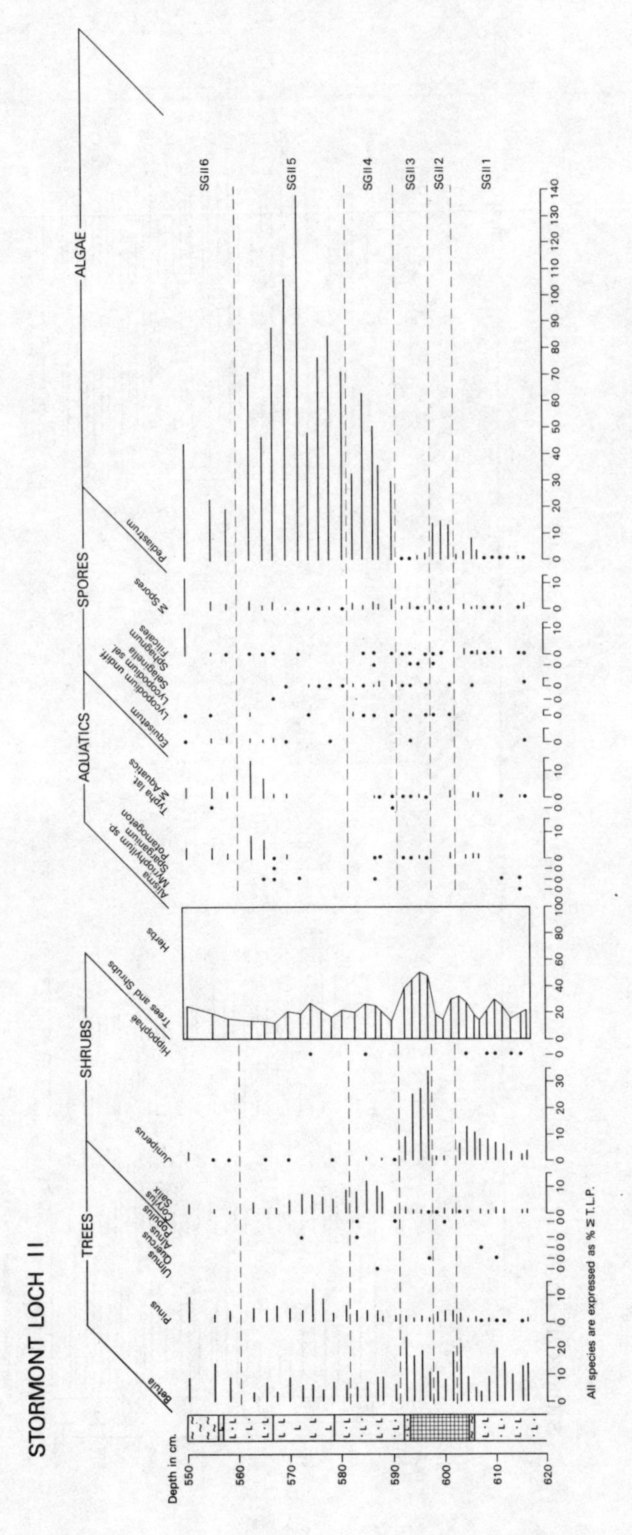

Fig. 3a Relative pollen diagram for SGII showing trees, shrubs, aquatics, spores and algae.

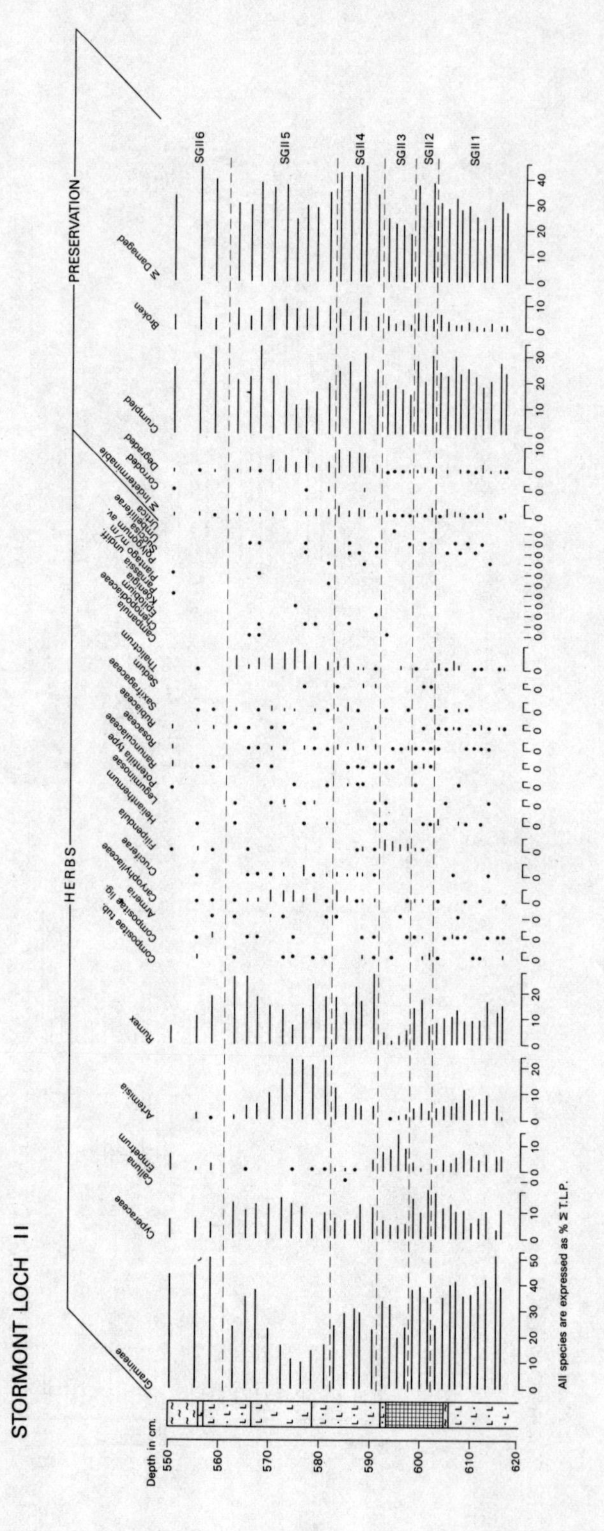

Fig. 3b Relative pollen diagram for SGII showing herbs and preservation.

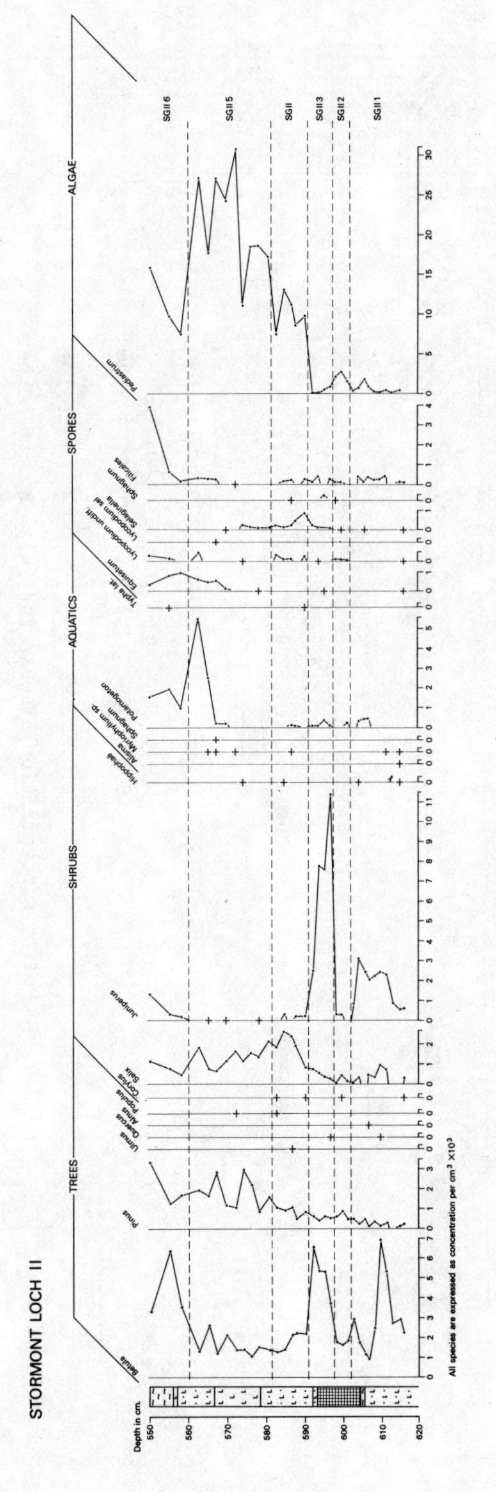

Fig. 4a Pollen concentration diagram for SGII showing trees, shrubs, aquatics, spores and algae.

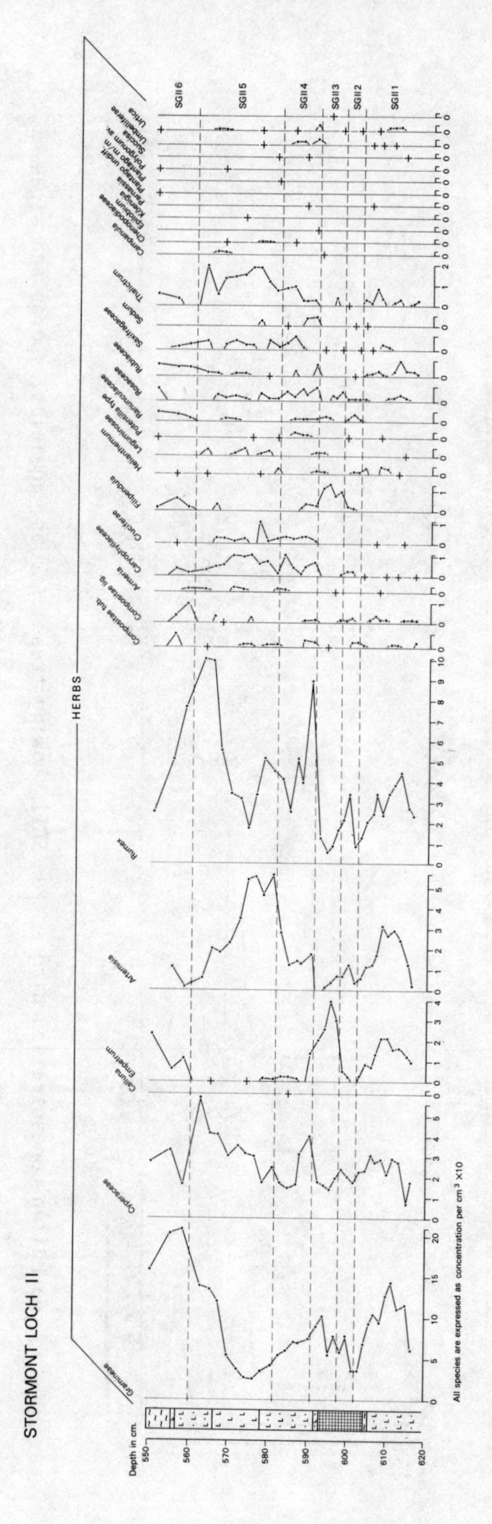

Fig. 4b Pollen concentration diagram for SGII showing herbs.

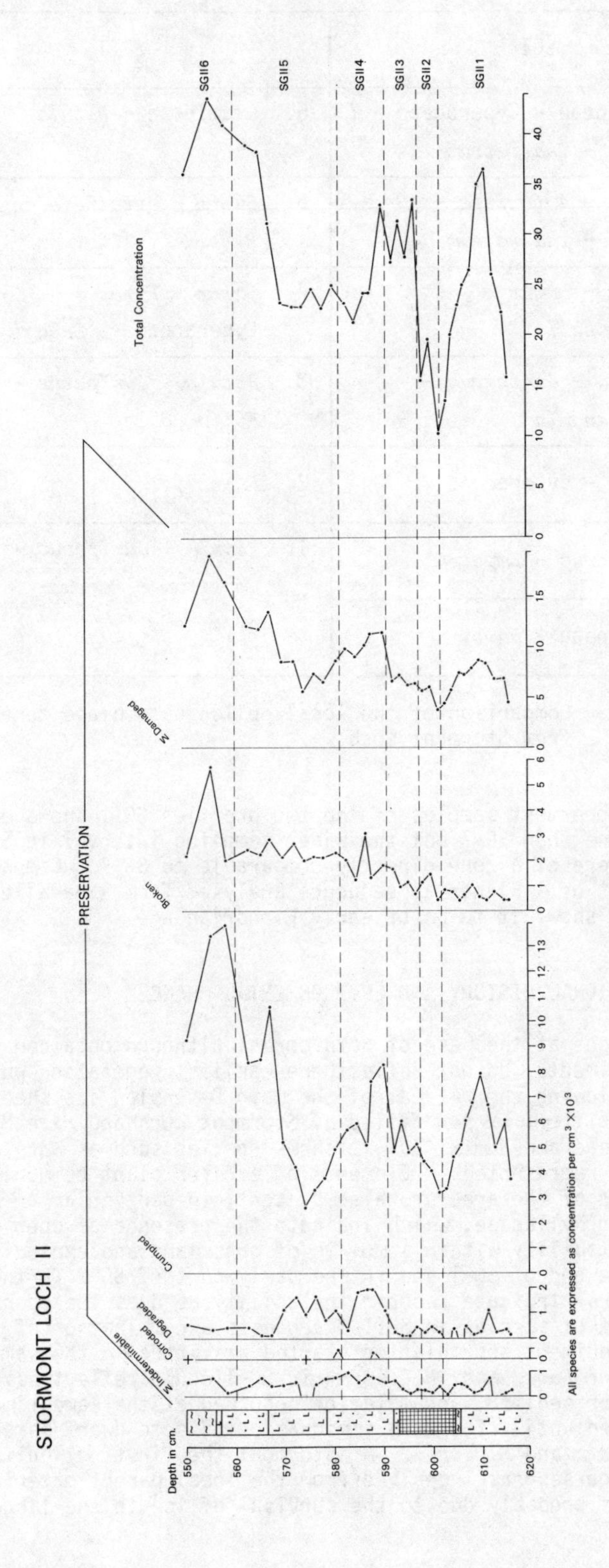

Fig. 4C. Pollen concentration diagram for SGII showing pollen preservation and total pollen concentration.

SGI	SGII
7. Gramineae - Cyperaceae - *Rumex* - *Thalictrum*	6. Gramineae - *Betula*
6. *Salix* - *Artemisia* - *Rumex* - *Thalictrum*	5. *Salix* - *Artemisia* - *Rumex* - *Thalictrum*
5. *Salix* - *Rumex* - *Selaginella*	4. *Salix* - *Rumex* - Cyperaceae - *Selaginella*
4. *Betula* - *Juniperus* - *Filipendula*	3. *Betula* - *Juniperus* - *Filipendula*
3. *Rumex* - Cyperaceae	2. *Rumex*
2. *Juniperus* - *Empetrum*	1. *Betula* - *Juniperus* - *Empetrum* - *Rumex* - *Artemisia*
1. Gramineae - *Rumex*	

TABLE 1 Comparison of the local pollen assemblage zones from Stormont Loch

and SG-2. In the uppermost samples of the two profiles SGII shows evidence of a later assemblage zone than SG-7 but the wider sampling interval in SGII did not allow definition there of a zone directly comparable to SG-7. Comparison of SGII-6 with the lowest part of a Flandrian sequence analysed from a parallel core to SGI (Caseldine, unpub.) shows it to be of early Flandrian age.

VEGETATIONAL HISTORY AND ENVIRONMENTAL CHANGE

The pollen assemblages at the base of both cores, although obtained from predominantly inorganic sediments, do not reflect the earliest vegetation which developed around the loch following the retreat of the main Devensian ice sheet, the wastage of which left the kettle-holes now filled by Stormont Loch and Hare Myre. SG-1 is dominated by Gramineae and *Rumex*, but it shows species such as *Betula*, *Juniperus* and *Empetrum* to be well represented. Elements of earlier plant communities from the initial colonisation of the area are also present, in particular *Artemisia*, *Thalictrum* and Caryophyllaceae, which indicate the presence of open ground still subject to soil instability within a mosaic of grassland and expanding dwarf shrub heath. Towards the end of SG-1 and in the early part of SG-2 further increases in *Empetrum* and *Juniperus* indicate reduced instability as does the change to organic sediments. The mix of species in SGII-1 suggests a combination of SG-1 and SG-2 but at both sites sediment accumulation started at virtually the same time although the higher values for *Rumex* and *Thalictrum* in SG-1 could reflect slightly earlier deposition. Whether sediment accumulation occurred at the same time at both sites or not it was delayed until after the immigration of some dwarf shrub species, especially *Betula nana* and *Empetrum*. At Stormont the first polleniferous sediments lie directly over coarse gravel derived from the local parent material and the delay in sedimentation was probably due to the survival of ice in the large kettle-holes.

The mix of communities indicated in SG-1 and SGII-1 shows the prolongation effect described by Pennington (1977a) whereby disturbed habitats suitable for such species as *Rumex, Artemisia* and *Thalictrum* persisted well into the interstadial until after the immigration of *Juniperus* and *Empetrum*. This effect would have been heightened by the presence of stagnating ice in the kettle-holes and continued instability around the margins of the depression.

The Gramineae - *Rumex* assemblage zone found in SGI shows affinities with the *Rumex* - Gramineae assemblage zone dated to c. 13,000 - 13,800 BP at Windermere except for the lower proportion of thermophilous species present in this more northerly area. This zone is found at several sites throughout northern Britain, especially at the base of Lateglacial profiles, and in the Grampians is found in B1 at Blackness (Walker, 1975a) and C1b at Cambusbeg (Lowe and Walker, 1977), but Stormont shows lower *Salix* values than in these equivalent zones. Pennington has commented that "*Salix* values in the Younger Dryas deposits of Britain are usually much lower than in early Late Devensian deposits" (1977b p.264), but the reverse appears true here. Low *Salix* values in a Gramineae and *Rumex* dominated zone are also characteristic of Loch of Park and Loch Kinord in 'Zone I' (Vasari and Vasari, 1968; Vasari, 1977). It would appear that the absence of high *Salix* values in the Lateglacial Interstadial is restricted to the eastern and southern Grampians for at several sites in this area *Salix* values are consistently higher in the Loch Lomond Stadial (*e.g.* Blackness, Roineach Mhor and Amulree - Lowe and Walker, 1977). The variability in *Salix* values may only represent a difference in the location of sites relative to the nearest *Salix* communities for its pollen is poorly dispersed but a distinctive pattern appears to be emerging which could be related to specific environmental parameters.

The early vegetation development phase (Lowe and Walker, 1977) represented in the lower levels of both cores is also marked at the base of SGII-1 by low pollen concentration values of only 15×10^3 grains/cm^3 but this gradually increases through the zone to over 36×10^3 grains/cm^3 at 611 cm (Fig. 4c). In the relative diagram there is the increased dominance of woody species, especially *Juniperus* and *Empetrum*, and at SGI this marks the boundary between SG-1 and SG-2 with, shortly after, the change from inorganic to organic sediments. At the deeper site, SGII, the change to organic mud occurs slightly later. It would appear from the percentage counts that the local expansion of *Juniperus* and *Empetrum* took place to the detriment of Gramineae and Cyperaceae but the concentration figures show that as the shrub species increased their representation so did all other species. What is more, the peaks in pollen concentration for *Betula* and *Empetrum* occur at the same time as the peaks for *Artemisia, Rumex* and Gramineae although *Juniperus* retains its values for a longer period. This may reflect a period of generally higher pollen production or could be due to a slower sedimentation rate as soils became more stable but indicates that there was no substitution of different communities actively taking place. The continued presence of a diverse vegetation cover including open ground dominated by *Artemisia, Rumex, Thalictrum*, Compositae (Liguliflorae) and Caryophyllaceae, grassland and dwarf shrub heath dominated by *Empetrum* and *Juniperus* with *Betula* (probably mainly dwarf birch on the basis of mean diameters) is not unusual in the early interstadial, especially at northern latitudes. The increased concentration of damaged pollen is not thought to be representative of any mixing of inwashed older sediment but is due to the higher general pollen concentration for there is little change in the relative amount of damaged pollen between SG-1 and SG-2 or in the later samples of SGII-1.

Most sites in Scotland show rises in *Betula, Juniperus* and *Empetrum* at this stage of vegetation development but very few show such changes as an isolated peak, especially in the Grampians. The general trend at Stormont closely parallels changes in the early interstadial represented by Zone A2 at Loch Sionascaig, Loch Tarff, Lochan an Smuraich and Cam Loch (Pennington *et al.*, 1972; Pennington, 1975a), except for lower frequencies of *Empetrum* at Stormont but *Empetrum* characteristically shows higher percentages in northern and western areas with higher precipitation and more oceanic conditions (Brown, 1971; Walker, 1975b). Similar early peaks for *Juniperus* appear in north-east Scotland at Loch of Park, Loch Kinord and Garrall Hill (Vasari, 1977) and in southern Scotland at Corstorphine (Newey, 1970). Such a pattern is also found in other British sites in the Lake District (Oldfield, 1960; Pennington, 1970) and in Yorkshire (Bartley, 1962). There is however a noticeable lack of comparable sequences in the Grampians although at Tynaspirit and Tirinie (Lowe and Walker, 1977) there are double peaks for *Juniperus* in the interstadial record. Zone C2, a *Betula* - *Juniperus* - *Empetrum* assemblage zone, found at Corrydon (Walker, 1977) is the only zone similar to SG-2 except for slightly higher *Empetrum*, and this zone Walker equates with the Bölling interstadial or 'Zone Ib'.

The gradual development of a more closed vegetation cover with the extension of shrub species was disturbed by a distinct revertence in the vegetational sequence recorded in SG-3 and SGII-2 by the decline and disappearance of *Juniperus* and *Empetrum* and reduced frequencies for *Betula*. This change appears sudden in the diagrams based on relative frequencies but pollen concentration had been falling before the sharp changes in the percentage curves. During SG-3 and SGII-2 the only species to increase their pollen representation are *Rumex*, and to a much lesser extent *Artemisia*. Overall the zones are characterised by very low pollen concentration, even lower than in the later stadial. Areas of dwarf shrub heath dominated by *Empetrum* and *Juniperus* must have disappeared and any scattered tree birches would also have been made unable to survive. The change in the vegetation cover did not lead to any widespread increase in grassland or any return to unstable conditions for values of Gramineae do not reach those experienced earlier and the rise in Cyperaceae seen in the relative pollen diagrams is not mirrored in the concentration figures. There is no change in the lithostratigraphy despite the small increase in *Artemisia* which would be taken as evidence for a resumption of solifluction.

From their pollen assemblage, low pollen concentration and stratigraphic position SG-3 and SGII-2 would appear to correlate with the Older Dryas chronozone (Mangerud *et al.*, 1974) which in northern Scotland is thought to be represented by the *Rumex* - *Artemisia* assemblage zone (Pennington, 1977b) and in northern England has been discovered at Blelham Bog (Pennington and Bonny, 1970) and Tadcaster (Bartley, 1962). At Corrydon the equivalent zone is C3, a *Selaginella* - *Rumex* - Compositae assemblage zone, which shows a greater degree of revertence in the vegetation communities towards more chionophilous and broken ground forms as would be expected at the higher altitude. At Loch of Park there is a similar increase in *Rumex* but little change in *Artemisia* and Vasari (1977) defines a 'Zone Ic'. Loch of Park and Corrydon remain the only other sites in eastern Scotland which show evidence for such a change in the development of vegetation during the interstadial. For all other sites in eastern Scotland, the facts support the conclusion offered by Gray and Lowe (1977) that "the vegetational succession and soil development progressed without interruption from about 13,000 BP until the beginning of the Loch Lomond Stadial" (p.170).

In view of the apparent contradiction of sequences within this part of eastern Scotland it may be that the results from Stormont do not necessarily represent genuine environmental change but are a function of sediment instability leading to the redeposition of older material and hence older pollen. This is thought unlikely on several grounds. There is no change in the lithostratigraphy in either core coincident with the apparent vegetation changes which would indicate slipped sediments or reworked basin edge deposits (Gray and Lowe, 1977). There is also little change in the concentration of damaged pollen when compared with zones above and below. The reduction in pollen concentration in SG-3 and SGII-2 is very marked with considerably lower values recorded than elsewhere in the diagram, due probably to a combination of reduced local pollen production and an increased sedimentation rate, and not to the techniques used, for the changes are greater than those that would be expected due to sampling error. The variability in concentration (Figs. 4a,b,c) also serves to emphasise the changes in the relative pollen diagrams.

Furthermore the species represented in SG-3 and SGII-2 do not appear to reflect a combination of older assemblages, they rather form a distinctive separate assemblage.

The small scale of the changes indicated at Stormont, as at Corrydon, is probably responsible for the apparent absence of similar features elsewhere in eastern Scotland. The variations in pollen frequencies occur over only 6 cm or less and with differing sedimentation rates and wider sampling intervals such a sequence may not be observed. This would be particularly true of sites where relative pollen diagrams show a great deal of variability in frequencies for individual taxa throughout the Lateglacial. It does nevertheless seem surprising that such a pattern should be evident at a lowland site and not at sites at greater altitude, with the exception of Corrydon, which may be expected to be more 'marginal' and sensitive to climatic change. If however vegetational development had not progressed sufficiently at higher altitudes to allow the establishment of dwarf shrub heath and in particular the local appearance of *Juniperus*, then a brief deterioration of climate, especially if there was no increased soil instability, may not have been sufficient to be registered in the pollen record and vegetational development may have been arrested for a short time rather than reversed.

The occurrence of a deterioration in climate at *c*. 12,000 BP has now been established in Britain on the evidence of both fossil pollen and coleoptera although the latter do not show any subsequent improvement in climate before the early Flandrian (Pennington, 1977a; Coope, 1977). In northern Scotland and in the Lake District Pennington has suggested that the recession marked by the 'Older Dryas' lasted only about 200 years between 12,000 and 11,800 BP and this short time scale, if valid for the eastern half of the country, could further account for the absence of any record in some deposits, especially where there was a slow accumulation rate.

Following this retrogressive phase of vegetation development there is an expansion of woody species with successive peaks for *Empetrum*, *Juniperus* and *Betula* at the expense of those species characteristic of open ground such as *Artemisia*, Caryophyllaceae, *Thalictrum* and Ranunculaceae. *Empetrum* reaches values of 12%, a concentration of 4×10^3 grains/cm^3, relative figures comparable to those found at Morrone in Aberdeenshire and Blackness (quoted in Birks and Mathewes, 1978) but much lower than at sites further to the north and west emphasising the pattern commented upon by Walker (1975b). In both diagrams *Juniperus* is the major pollen contributor producing twice as much pollen as *Betula* and whereas the concentration of *Betula* is similar to that in the earlier zones, the concentration of *Juniperus* is almost quadrupled, a difference which is not so clear in the relative figures (Figs. 2a and 3a). Herbaceous pollen is much reduced except for Gramineae but there is a peak for *Filipendula*. There is also some expansion of the aquatic flora with not only *Potamogeton* but also *Myriophyllum spicatum* and *M. alterniflorum* represented.

The character of the vegetation at this time was one of juniper-dominated heath with the local appearance of scattered copses of tree birches. Although there were still open areas of grassland with *Rumex*, few other herbaceous species were present and there is no evidence of any disturbed ground. Both zones SG-4 and SGII-3 which show this vegetation record appear typical of the later phase of the Lateglacial Interstadial as found throughout Britain, usually as a *Betula* - *Juniperus* pollen assemblage zone (Pennington, 1977a) and which is dated to between 11,800 and 11,000 BP. In the Grampians the *Betula* - *Juniperus* - *Filipendula* assemblage zone at Stormont can be correlated with T2c at Tynaspirit 2, B4 at Blackness and C1d at Cambusbeg (Lowe and Walker, 1977). It differs from most of the northern sites in having higher *Betula* values, just as it differs from the main interstadial zone at Corrydon, C4, where neither *Betula* nor *Juniperus* achieves more than 15% of the count. Walker considers that Corrydon lay above the zone of birch copses for most of the pollen there is referable to *Betula nana*. Thus tree birches must have been restricted to Strathmore and the lower end of the glens as seen from the diagrams for Blackness and Roineach Mhor with *Juniperus* locally abundant at lower altitudes (Lowe and Walker, 1977). It is during this period that the record at Stormont most closely correlates with Loch of Park and Loch Kinord (Vasari, 1977) but there is only one sample which could be assigned to a later *Juniperus* peak as recognised at Loch Kinord and also Abernethy (Birks and Mathewes, 1978).

The deterioration in climate which allowed the redevelopment of ice masses in the Grampians (Sissons, 1974a) is marked by changes in both the lithostratigraphy and the pollen record. SG-5 and SGII-4, in contrast to the preceding zones, reflect considerable disturbance of soils around the basin with a return to inorganic deposition, and, within the silt clays of SGI, bands of redeposited organic material. The woody species dominant during the interstadial are severely reduced and there is an expansion of those species found earlier in the pollen record, especially *Rumex*, Caryophyllaceae, Ranunculaceae, *Thalictrum* and *Artemisia*. Although some changes are sudden, such as the drop in concentrations of *Betula* and *Juniperus*, others are gradual and after an initial reduction in pollen concentration to 25 x 10^3 grains/cm^3 values remain consistently between 20 and 25 x 10^3/cm^3. The early phase of the stadial is represented in SG-5 and SGII-4 by a rise in *Salix* but overall appears very much as a transition between interstadial and true stadial assemblages. Despite the low *Artemisia* counts relative to later zones, soil instability is clearly indicated in the lithostratigraphy and in the presence of *Selaginella selaginoides*, which was probably growing on freshly exposed base-rich rills and flushes around the loch (Birks and Mathewes, 1978). The increased inwash of minerogenic material is also seen in the rise in damaged pollen, particularly in the Degraded/Amorphous and Crumpled categories.

Due to the break up of soils, grassland communities could not expand to the same extent as in the pre-interstadial and the greater species diversity and higher Cyperaceae values suggest analogies with present day Alpine summit vegetation, such as the Cariceto - Rhacomitretum lanuginosi community (Birks, 1973). It is however difficult fully to accept such an analogy in view of the *Salix* values at Stormont. High counts for both *Salix* and Cyperaceae are found at Loch of Park, Corrydon, Roineach Mhor and Tynaspirit and appear to be restricted to a southern and eastern distribution. Interpretation in terms of known modern pollen assemblages however remains uncertain.

Because of the transitory nature of the assemblage zones SG-5 and SGII-4, few direct comparisons can be made with assemblage zones elsewhere in Scotland. At many sites deposits covering the stadial are represented by a single relatively uniform pollen assemblage (Pennington, 1977a), or by no substantial pollen record at all, especially close to areas of ice accumulation. SG-5 and SGII-4 do, however, show similarities with the *Rumex* - *Selaginella* assemblage zone at Tirinie, the Cyperaceae - *Rumex* - *Lycopodium* assemblage zone at Amulree 2 (Lowe and Walker, 1977) and the *Lycopodium* - *Selaginella* - Compositae - Cyperaceae assemblage zone at Corrydon (Walker, 1977). At all these sites there are low *Artemisia* values relative to later zones.

SG-6 and SGII-5 show a further increase in Cyperaceae and herbaceous species with maximum values for *Artemisia*, up to 20% and 6 x 10^3 grains/cm^3. Caryophyllaceae, Chenopodiaceae, Saxifragaceae, Cruciferae and *Thalictrum* are also well represented, the latter up to 2 x 10^3 grains/cm^3. This increase in species indicative of open ground is also seen in the greater variety of taxa for within the Caryophyllaceae there are grains referable to *Silene* type and *Cerastium alpinum* type, and within the Saxifragaceae there is a dominance of *Saxifraga aizoides/stellaris* type. Thus there was through this period a further opening of the vegetation cover with the most severe movement of soils due to solifluction. The consistently high contribution of *Pinus* pollen is not unusual and serves to emphasise the openness of the environment allowing greater representation of pollen transported from long distances.

Concerning the importance of *Artemisia* Birks and Mathewes (1978) and Walker (1975a) have shown how, in eastern Scotland and the Cairngorm-Grampian massif, frequencies for *Artemisia* tend to increase towards the Cairngorms. The difficulties associated with the interpretation of the significance of *Artemisia* frequencies are centred upon the lack of present day analogies and the known preference of *Artemisia* for xeric, well drained environments (Iversen, 1954) and dislike of long snow lie (Andersen, 1961). The problem is to reconcile these known ecological properties with the period of rapid glacier development during the Loch Lomond Stadial. Birks and Mathewes (1978) explain the high Speyside frequencies as the result of a rain shadow effect reducing precipitation from southerly and easterly winds, a feature also indicated from a reconstruction of firn lines (Sissons and Sutherland, 1976), therefore implying a higher precipitation level for the southern and eastern Grampians. This idea is supported by the lower *Artemisia* frequencies at Roineach Mhor and Blackness (Walker, 1975a). At the present time Blairgowrie lies within a rain shadow area and experiences much lower rainfall than within the Grampians but the values of 20% for *Artemisia* in the stadial are still high compared to all other sites. At most sites along the southern Grampians the main peak in *Artemisia* pollen occurs towards the middle or end of the stadial and at Stormont, Tirinie, Amulree and Corrydon this feature allows the definition of more than one assemblage zone to cover this period. It seems likely that, at least in the area covered by these sites, the delayed rise of *Artemisia* reflects an early period of high precipitation, characterised by higher *Salix* and lower *Artemisia*, which accompanied the main depression in temperature, and which was followed by a much drier phase once the ice had developed. The decrease in precipitation during the stadial coupled with the removal of the interstadial soil cover around Stormont Loch would then have produced conditions suitable for the local importance of *Artemisia*.

Towards the end of the stadial there is a sequence of falling *Artemisia* and rising *Rumex* followed in turn by an increase in Gramineae and the disappearance of open ground species. This change to a closed grassland cover precedes the appearance of *Empetrum* and *Juniperus* and the rise in *Betula* values but it is marked by a general increase in pollen concentration to values higher than those found during the rest of the stadial. The increase in pollen concentration may reflect the slower rate of sedimentation following the earlier period of soil instability but the change in species representation indicates changes in the local environment which could have resulted in greater pollen productivity although dating of this change would be difficult due to the inorganic nature of the sediment. It is possible that this change may be equivalent to the herb dominated assemblage zones found in northern Scotland after 10,500 BP for the lithostratigraphy does show signs of increasing organic content as periglacial processes became less effective. No comparable zone to SG-6 has however been defined elsewhere in the Grampians.

Although there is a reduction in the major woody species at the top of the SGII diagram, analysis of the overlying sediments elsewhere in the basin shows the expansion of birch woodland taking place soon after the change from inorganic to organic sediments. The transition between the Lateglacial and the early Flandrian, which probably occurred between 10,300 and 10,000 BP, is marked by a gradual change from open grassland to birch woodland through dwarf shrub heath dominated first by *Empetrum* and later by *Juniperus*, but the latter never expanded to the same extent in the early Flandrian as it did during the interstadial.

CONCLUSIONS

The importance of the results from Stormont Loch lies in the support which they give to the occurrence of a phase of retrogressive vegetation development prior to the Loch Lomond Stadial which may correlate with the Older Dryas of the north-west European stratigraphy, as proposed by Walker at Corrydon, but which has yet to be found anywhere else in the southern and eastern Grampians. Within the Loch Lomond

Stadial the definition of a series of assemblage zones allows closer identification of the sequence of vegetation change outside those areas immediately affected by the presence of glaciers, a sequence which indicates a reduction in precipitation after the initial period of ice development. The comparability of the relative pollen record and figures for pollen concentration further emphasises that the changes identified are genuine as does the comparability of the two relative pollen diagrams despite the local differences in lithostratigraphy.

ACKNOWLEDGEMENTS

Thanks are due to Lord Lansdowne for permission to work on the loch. The work was carried out while the author was in receipt of a University of St. Andrews Research Scholarship. Dr. Jack Jarvis provided valuable assistance in the field and most of the diagrams were prepared in the cartography unit of the Department of Geography, University of Exeter.

Environmental Change During the Loch Lomond Stadial: Evidence from a Site in the Upper Spey Valley, Scotland

J. B. Macpherson

(Memorial University of Newfoundland)

ABSTRACT

A kettle-hole site in the upper Spey valley reveals sediments which are ascribed to the Loch Lomond Stadial on both palynological and sedimentological grounds. Changes in the relative contributions of taxa suggesting oceanic and continental conditions, together with changes in the probable conditions of lacustrine sedimentation, permit a tentative reconstruction of a sequence of environmental phases. During the early part of the stadial snow-beds were probably extensive, since taxa considered to be chionophilous made significant contributions to the pollen spectra. Mineral sedimentation involved at first fine and then coarser particles. It is inferred that at first some summers were warm enough to permit thermal stratification of the kettle lake, but later cooler and perhaps windier summers prevailed. The most extreme conditions of the stadial, represented by the Artemisia *maximum, were marked by a probable reduction in snow-beds, a decrease in the rate of sedimentation and in the size of mineral particles, and a fine lamination of sediment which suggests cold monomictic conditions in the lake. Examination of* Artemisia *maximum pollen spectra from elsewhere in the Scottish Highlands suggests that continental conditions prevailed at many valley sites east and north of the main watershed, while sites to the west and south continued to receive appreciable snowfall. Oceanic conditions returned to the kettle site towards the close of the stadial; snow-beds appear to have become more extensive, more sediment entered the lake and more active summer circulation prevented the preservation of laminations.*

INTRODUCTION

In recent years the results of several palynological investigations in or adjacent to the upper Spey valley have been published, and a general pattern of vegetational changes in Lateglacial and Flandrian time has emerged (Birks, 1970; O'Sullivan, 1974, 1975, 1976; Walker, 1975b; Birks and Mathewes, 1978). This paper presents the results of an investigation of a site on the floor of the Spey valley, between the Truim confluence and the Lochain Uvie, and about 3 km north of Walker's site at Loch Etteridge (Fig. 1; Fig. 3). The site was selected in the anticipation that it might provide evidence of hydrologic conditions in the Spey valley during the eastward discharge of the Glen Roy - Loch Laggan proglacial lakes at the time of the Loch Lomond Stadial (Macpherson, 1978). Although the site provides only limited, negative evidence of the stadial river level, the stratigraphy does suggest the events which have shaped this section of the floor of the Spey valley during the Flandrian.

A pollen profile from the basal sediments may be related to the general vegetational sequence in the area, and spans the final episode of the Lateglacial and the early stages of the Flandrian. The site's particular characteristics (the contrasting habitats in its catchment and the greater depth of sediment than in other sites from the area) permit a detailed examination of environmental changes during the stadial. The sequence of changes which emerges is in agreement with Mitchell's hypothetical "frost cycle" (1977), developed for Ireland, suggesting that it may be of wider applicability.

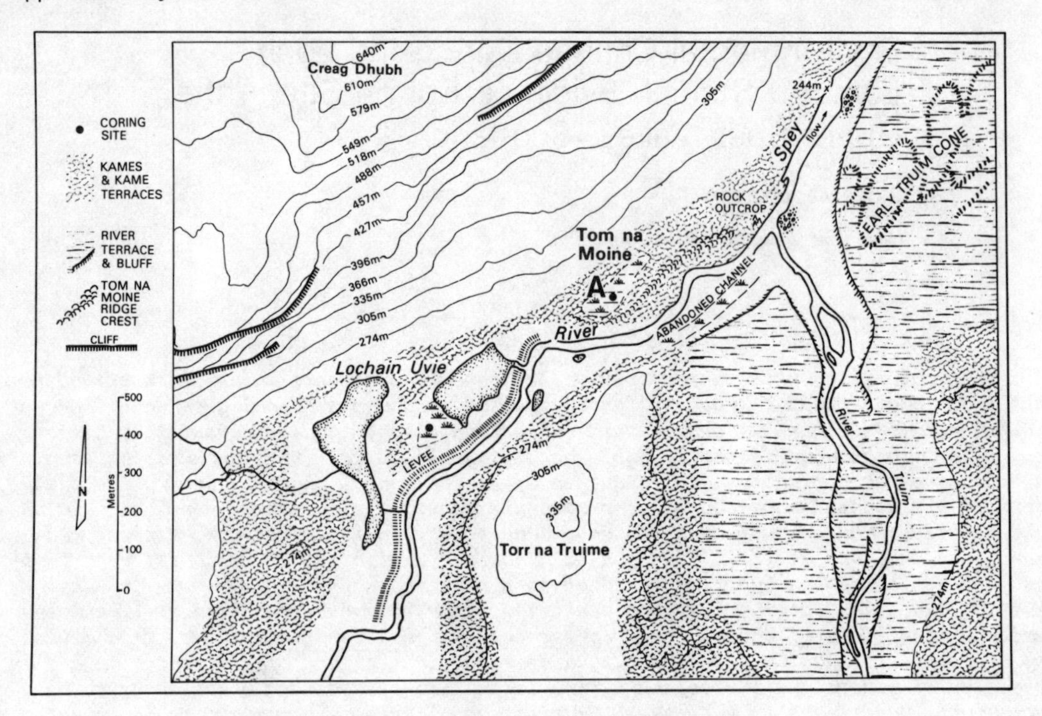

Fig. 1 Geomorphological features in the vicinity of the Tom na Moine site. General location of the site indicated on Fig. 3.

THE SITE

The coring site (Fig. 1, point A) lies at the deepest point of a bog (Nat. Grid. Ref. NN 683961) filling a kettle depression in deposits of the Late Devensian ice sheet (Sissons and Walker, 1974) between the River Spey and the hill known as Creag Dhubh (*c.*745 m). The bog is oval in shape, being about 250 m in length from SW to NE, and about 100 m wide. No stream drains into the bog today, nor is there any gully on the slope of Creag Dhubh of sufficient size to suggest that the depression has received the discharge of a perennial stream in the past. The kettle is separated from the Spey flood-plain by a ridge of coarse gravel known as Tom na Moine ("the hill of the bog") which rises 14 m above the bog at its maximum, but which is only about one metre high at the SW end of the bog, where a drainage ditch has been cut through it. The level of the bog surface (*c.* 243 m above sea-level) is the same as that of the flood-plain, suggesting that the water-table in both is continuous, the coarse gravel of the buried part of the ridge forming no significant hydrologic barrier.

Two attempts were made to core the site. In the summer of 1973 a Hiller sampler was used to penetrate 10.85 m of peat and underlying lacustrine sediments. At several levels the peat was so fluid that it could not be retrieved, and the lacustrine sediments proved to be contaminated. A second attempt was made with a piston corer in November of that year. At that time the peat was even wetter, but it was possible to obtain a core for pollen analysis from the more cohesive basal sediments between 9.65 m and 12.15 m.

A generalized stratigraphic profile is as follows:

with Hiller sampler:

0 - 8.50 m	*Sphagnum* peat, poorly humified to 1.75 m, well humified and very wet at depth; wood encountered at 3.25, 3.40, 4.30-4.40 m
8.50 - 9.50 m	very wet sedge peat
	(hiatus)

with Dachnowski corer:

9.65 - 10.00 m	fibrous gyttja
10.00 - 10.45 m	gyttja
10.45 - 10.68 m	organic mud
10.68 - 11.15 m	silty clay
11.15 - 11.25 m	laminated silty clay
11.25 - 11.80 m	silty clay with grit particles
11.80 - 12.15 m	clay with organic-stained layers at 11.85, 11.92, 11.94, 12.13 m
12.15 m	base, apparently gravel.

There is no stratigraphic evidence that Spey water has ever entered the depression directly; in other words the Tom na Moine ridge has always functioned as a barrier between the kettle depression and the river. It will be noted that the upper level of the sedge peat is 8.50 m below the present level of the Spey flood-plain. The upper 8.50 m of bog peat at the site could not have accumulated with the Spey and its flood-plain at their present level, and it is evident that aggradation by the Spey has occurred while the bog peat has accumulated.

The site lies at the lower end of a reach of the Spey valley floor which is so flat that earlier workers (Jamieson, 1865, 1892; Hinxman, 1901) considered it to be the floor of a former lake. The stratigraphic evidence presented here suggests, on the contrary, that the very low gradient of the flood-plain is the result of fluvial aggradation, the cause of which can only be the deposition in the Spey valley of the cone of coarse gravel which is evident at the mouth of its tributary, the Truim. It is of significance that while erosional terraces occur in the Spey valley immediately below the Truim confluence, there are none for several kilometres upstream. Young (1978) also comes to the conclusion that this reach of the Spey valley floor is of fluvial rather than of lacustrine origin.

The stratigraphy of the Tom na Moine site indicates that sedimentation began while slope processes in the basin yielded mainly mineral sediment. Two components of this sediment are unusual: the organic-stained layers near the base and the finely laminated silty clay between 11.15 m and 11.25 m. These will be discussed further in relation to the evidence they provide of environmental conditions. The minerogenic sediment is overlain by organic clay and then gyttja; this transition is typical of the change from the unstable soil conditions of the Loch Lomond Stadial to the increasingly stable soils of the early Flandrian, a correlation which is confirmed by pollen analysis (Fig. 2). The change in sediment from gyttja to sedge peat and finally to *Sphagnum* peat indicates a normal hydroseral succession.

POLLEN ANALYTICAL METHODS

Samples for pollen analysis were taken from the core at intervals which were suggested by the stratigraphy. It is recognized now that a closer sampling interval would have been useful in certain parts of the core. The samples were processed by Erdtman's acetolysis method, preceded where necessary by treatment with hydrofluoric acid. Pollen was scarce in the basal mineral sediment, but a count of at least 100 land pollen grains was made from every sample except that from 11.94 m. The number of slides counted for each sample is indicated in Fig. 2. In discussing vegetational changes during the stadial the counts were pooled according to the sediment from which the samples were taken, giving, in general, a larger and more reliable pollen sum. Samples from the more organic sediment above 11.68 m were richer in pollen; mineral counts of 150 tree pollen grains and 225 land pollen grains were attained or exceeded at each level. The percentage base for the pollen profiles is the sum of land pollen grains; the percentages of aquatic pollen and of spores are calculated on the same base.

The pollen profile was zoned by subjective recognition of pollen assemblages. Gordon and Birks (1972a) and Pennington and Sackin (1975) indicate that the boundaries of zones drawn in this manner are likely to correspond with boundaries obtained by numerical analysis. The work of Pennington and Sackin, moreover, reveals a close correspondence between pollen zone boundaries and environmental changes indicated by chemical analysis. The writer feels justified, therefore, in making correlations between this site and others in the same general area of the Highlands on the basis of pollen assemblage zones, despite the lack of radiocarbon dates.

POLLEN ZONATION

The pollen assemblage zones which are recognized in the Tom na Moine profile (Fig. 2) are listed in Table 1. The successive maxima in *Empetrum*, *Juniperus*, and *Betula* in zones TM-3, TM-4 and TM-5 occur regularly at the transition from the Lateglacial to the Flandrian in Scotland (Vasari, 1977). The table also indicates suggested correlations with the three other sites in the area of the Spey valley which have revealed Lateglacial assemblages; there is good correspondence between the zonations of the four sites. Radiocarbon dates from the Loch Etteridge site are indicated in the Table (Sissons and Walker, 1974; Walker, 1975b), together with the interpolated radiocarbon ages of the zone boundaries from the Abernethy Forest site (Birks and Mathewes, 1978). While there are discrepancies between the two sets of dates, they leave little doubt that the period of vegetational revertence, of which evidence is seen in the *Artemisia* pollen assemblage zones, LE-5 and AFP-3, should be correlated with the Loch Lomond Stadial.

It will be noted that a more detailed pollen zonation is given for the Tom na Moine site than has been suggested for the other sites. A much greater depth of minerogenic sediment accumulated in this small deep kettle depression than at the other sites, and this has allowed greater resolution of the profile. Since the Lateglacial - early Flandrian sequence of vegetational changes has already been described (Birks,

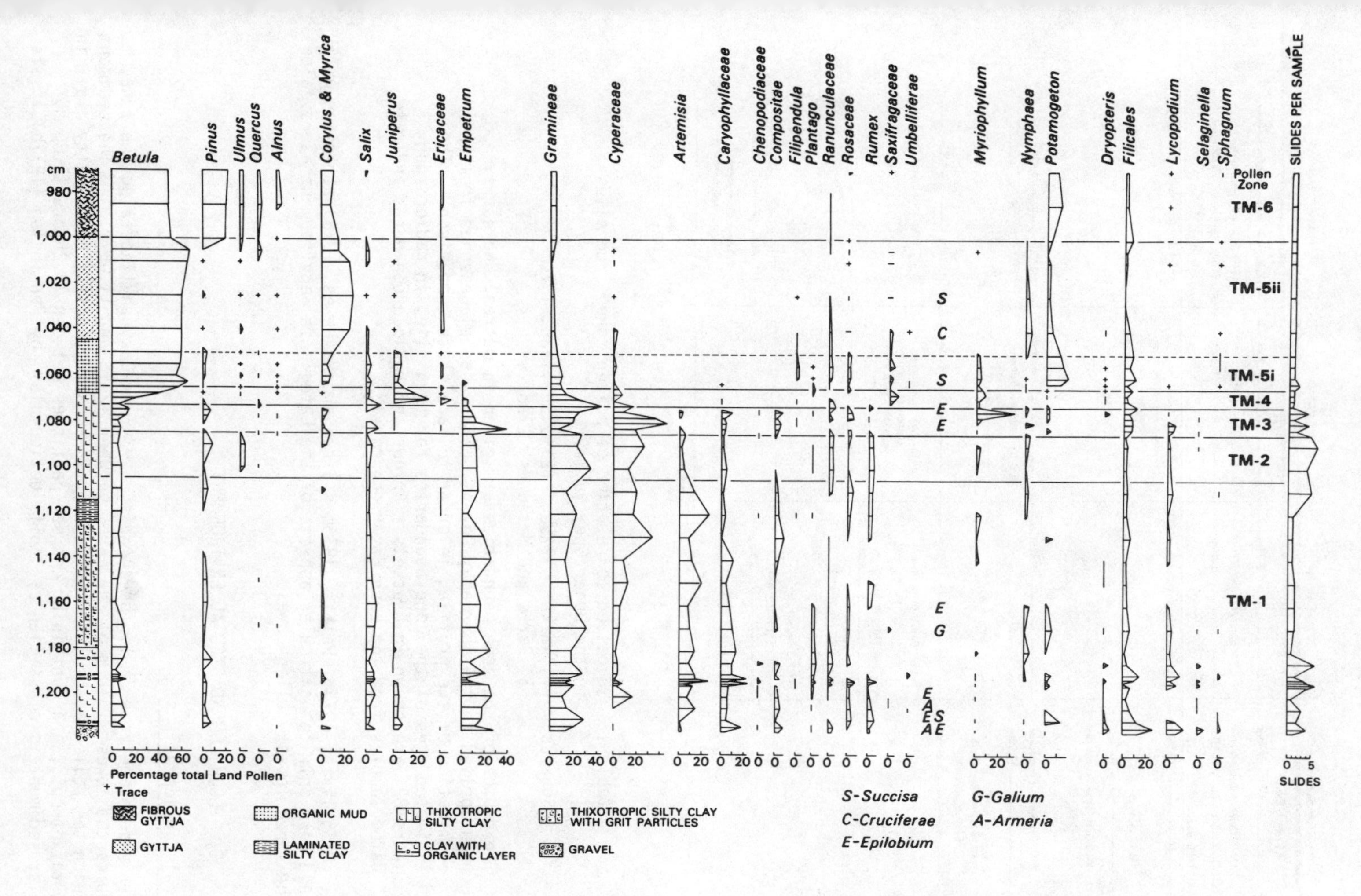

Fig. 2 Pollen profile from the lower part of the Tom na Moine stratigraphy

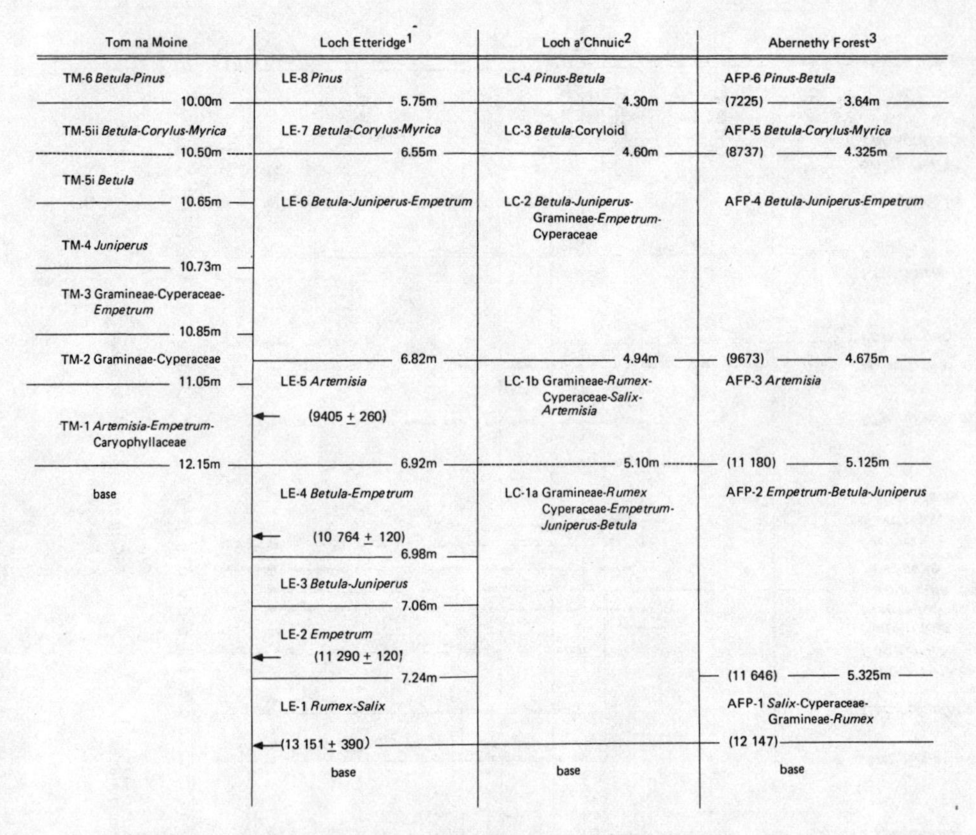

Table 1 Suggested correlation of pollen zones from sites in the Upper Spey valley

Sources: (1) Walker, 1975b; (2) O'Sullivan, 1974; (3) Birks and Mathewes, 1978

The depths of the zone boundaries are given in metres. Names have been applied to the Abernethy Forest pollen assemblage zones from an examination of the profile, in order to facilitate comparison. The estimated ages of the zone boundaries for this site, in radiocarbon years BP, are shown in brackets. Arrows and radiocarbon dates indicate dated samples from the Loch Etteridge core.

1970; O'Sullivan, 1974; Walker, 1975b; Birks and Mathewes, 1978) attention will be paid here only to particular aspects of the Lateglacial environment as revealed by the three lowest pollen zones.

LOCH LOMOND STADIAL ENVIRONMENT

General Considerations

Pollen assemblage zone TM-1, the *Artemisia-Empetrum*-Caryophyllaceae pollen zone, occupies the lowest 1.10 m of the core, between 12.15 and 11.05 m. The sediment in which the pollen accumulated is mainly minerogenic, consisting of silty clay, in part thixotropic, and containing coarser particles between 11.25 and 11.80 m. Similar sediment is classified as "solifluction clay" by Pennington (1977b). Its source in the Tom na Moine depression was the Late Devensian glacial deposits surrounding the basin.

These glacial deposits probably underlie the basin, and were presumably encountered in the impenetrable gravel at the base of the core. The absence of sediments containing Lateglacial Interstadial pollen assemblages may perhaps be attributed to free drainage through the gravel with a lower water-table in the absence of permafrost. *Juniperus* pollen is continuously present at the base of the core and the proportion of *Artemisia* is very small in the three lowest samples. Since juniper, among other woody taxa, was present in the interstadial vegetation at the other Spey valley sites, it is possible that the lowest part of pollen zone TM-1 represents the disruption of the interstadial vegetation and its replacement by vegetation adapted to the harsh conditions of the stadial.

Habitats in the Catchment

While zone TM-1 is named for the distinctive pollen of the assemblage, namely *Artemisia* (1-26%), *Empetrum* (5-28%) and Caryophyllaceae (1-22%), significant contributions are also made by Gramineae (11-34%), Cyperaceae (0-34%), *Betula* (3-13%), *Salix* (2-9%) and families of herbs which include species indicative of open, disturbed ground: Compositae, *Plantago* and *Rumex*. There are some marked percentage variations in most of these taxa which may well simply reflect the low pollen sums available from this site. Nevertheless, some tentative comparisons are suggested.

The percentage of *Artemisia*, although substantial, is generally smaller than at Loch Etteridge, where it reaches 30-40%, and at Abernethy Forest, where it exceeds 50%. At the same time, however, the percentage contribution made by *Empetrum* to the Tom na Moine spectra is considerably greater than at the other two sites, being similar to the percentage recorded for Loch Sionascaig, in NW Scotland (Pennington *et al.* 1972). Values for Caryophyllaceae are also higher than at Loch Etteridge and Abernethy Forest, reaching maxima in narrow bands of organic-stained silty clay near the base of the core, with a median value for the zone of 8%. Comparable percentages in stadial sediments are found elsewhere in the Highlands only at Loch Kinord in the Dee valley (Vasari and Vasari, 1968) and at the Roineach Mhor (Walker, 1975a) and Amulree (Lowe and Walker, 1977) sites in the southern Grampians. In Birks' analysis of modern Scottish pollen spectra (1973), only one site (Glen Douchary in Wester Ross) yielded as much as 10% Caryophyllaceae pollen, yet Caryophyllaceae (almost all *Silene acaulis*) constituted over 25% of the vegetation cover of the site. If these values are applied to the fossil pollen records from the Tom na Moine site, it is possible that Caryophyllaceae formed a very significant component of the stadial vegetation cover in the vicinity of the site.

The distinguishing topographic characteristic of the catchment is the slope of Creag Dhubh. Its mosaic of rocky ledges, cliffs and screes draining directly into the basin affords a wider range of habitats than occur at either Loch Etteridge or Abernethy Forest. Some estimate may be made of the permafrost and moisture conditions of these habitats during the stadial by considering climatic parameters derived from work on firn-line elevations. Sissons (1974a) calculated a mean July sea-level temperature of 7.6°C for the Gaick Plateau to the SE of the site, which with normal lapse rates would indicate a mean July temperature at the site of about 6°C. Coope (1977) suggests a mean annual range of temperature of 35°C, using evidence from fossil coleopteran assemblages from various parts of Britain. Such a temperature regime suggests scanty winter precipitation. Firn-line calculations for the SE Grampians and the NW Highlands suggest that while the outer margins of the Highlands received substantial snowfall, there was a rapid decrease in precipitation toward and beyond the watershed (Sissons and Sutherland, 1976; Sissons, 1977b). Indeed, Sissons (this volume) suggests an annual precipitation of only 200-300 mm in Speyside during the stadial.

There can be little doubt that the catchment was underlain by permafrost during the stadial. Its effects would have varied depending on soil drainage, available moisture and the thickness of the active layer. The irregular topography of the upper slopes of Creag Dhubh would have allowed the accumulation of snow-beds, but soil flow is unlikely to have been extensive because of the proximity of bedrock to the surface. The lower slopes of Creag Dhubh are mantled with sandy drift, and would have been supplied with moisture from the melting of snow from higher on the slope. The moisture-retaining capacity of the drift and the southerly aspect of the slope, promoting a deeper active layer, are likely to have resulted in soil flow. By contrast, the gravel ridge which separates the site from the Spey valley proper is likely to have been well drained even under conditions of permafrost (Rieger, 1974); its smooth slopes may have been largely swept clear of snow by wind. Here cryoturbation, rather than soil flow, is likely to have been dominant.

What may be a modern approximation of the stadial environment of the upper Spey valley occurs in the continental zone at the heads of the SW Greenland fjords. Precipitation is in the range 100-250 mm, and the snow has disappeared from the head of Søndre Strømfjord by the end of April (Böcher, 1954). However, the summer temperatures cited by Böcher are higher than those calculated for the Scottish Highlands; at the head of Søndre Strømfjord mean temperatures exceed 8°C for three months.

Although little is known of modern pollen assemblages from this region of Greenland, the vegetation includes plants similar to those which produced the characteristic pollen of pollen zone TM-1. *Artemisia* is abundantly represented by *A. borealis* and Pennington (1979) reports a pollen spectrum containing 40% *Artemisia* pollen from a moss polster near the head of Søndre Strømfjord. *A. borealis* is common where there is a light and early-melting snowcover, and tolerates a wide range of disturbed but well-drained sites from summit fell-fields and scree slopes to dunes and wind-eroded river plains (Böcher, 1952, 1954, 1963). In the Tom na Moine catchment corresponding habitats would have occurred on scree slopes on Creag Dhubh and on the gravel ridge closer to the basin. The Caryophyllaceae pollen grains encountered in the Tom na Moine samples were not differentiated as to species. However, plants of this family occur in sites in continental SW Greenland similar to those occupied by *Artemisia*.

Empetrum occurs in continental SW Greenland only as an early snow-bed or bog plant. It is widespread, by contrast, in the exposed oceanic heaths of the outer SW coast (Böcher, 1952, 1954, 1963). It has some tolerance for both structural and amorphous solifluction: Böcher reports that it will grow on patterned ground, presumably at the stable margins of active soil polygons as it does on exposed oceanic summits in Newfoundland (Day, 1978). Dahl (1956) indicates that it is somewhat resistant to amorphous solifluction in Rondane, Norway, but is seldom present in the more active solifluction soils.

Snow-beds on the upper slopes of Creag Dhubh could have provided sites enabling *Empetrum* to survive the harsh conditions of the stadial. Pennington *et al.* (1972) have suggested a secondary origin for at least some of the *Empetrum* grains encountered in stadial lake sediments from northern Scotland. It is possible that a proportion of the *Empetrum* grains in pollen assemblage zone TM-1 is of secondary origin, for many of the grains are crushed; however, the crushing could have occurred in transit from the slopes of Creag Dhubh if pollen from stadial snow-beds were washed into the basin by overland flow.

GEOGRAPHICAL VARIATION OF LOCH LOMOND STADIAL POLLEN SPECTRA

A consideration of the spatial variation in *Empetrum* pollen percentages lends support to the argument for its persistence in favourable locations throughout the stadial. Fig. 3 is an attempt to show the variation in published pollen spectra ascribed to the stadial for Highland Scotland, using the cartographic method of Godwin (1975). The horizon selected was that assumed to mark the most severe conditions of the stadial, the *Artemisia* maximum. Certain sites were excluded, either because percentages of *Artemisia* were so low that this horizon could not be defined (*e.g.* Birks' (1973) sites on Skye), or because the pollen sum used was not the sum of land pollen (*e.g.* O'Sullivan's (1974) Loch a'Chnuic site).

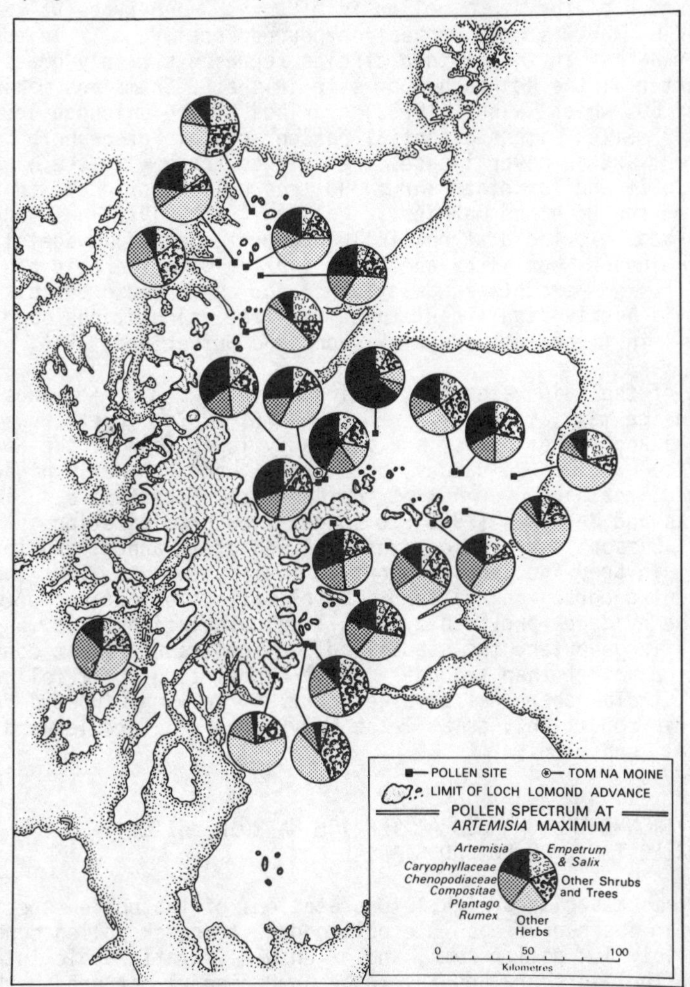

Fig. 3 Pollen spectra at the *Artemisia* maximum from sites in and adjacent to the Scottish Highlands

(Sources: Birks and Mathewes, 1978; Kirk and Godwin, 1963, redrawn in Birks, 1973; Lowe and Walker, 1977; Pennington, 1975b, 1977; Pennington *et al.*, 1972; Vasari, 1977; Vasari and Vasari, 1968; Walker, 1975a, 1975b, 1977).

At about a quarter of the mapped sites the percentage value for *Empetrum* is exceeded by that for *Salix* at the *Artemisia*-maximum horizon. At most of these sites the authors note the presence of *Salix herbacea* macrofossils or of *S. herbacea*-type pollen (Vasari and Vasari, 1968; Walker, 1975a, 1977; Rymer, 1977), and suggest the existence of snow-beds. It was considered appropriate, therefore, to group *Salix* with *Empetrum* in the diagrams, since together they probably indicate oceanic or snow-bed conditions. The sites with the highest percentages of *Empetrum* plus *Salix* pollen lie generally at the S and NW margins of the Highlands. This peripheral distribution becomes more apparent if the percentage values for the sum of other shrub and tree pollen is considered in conjunction with the sum of *Empetrum* and *Salix*. Doubtless a fraction of the tree pollen is exotic, but at some sites a significant proportion of the birch pollen is of *Betula nana* type (Walker, 1975a, 1975b; Pennington, 1977b) and was most probably produced locally. In addition, the category "other herbs" in the divided circles represents mainly grass and sedge pollen, the latter at the Roineach Mhor site in the SE Grampians making a notable contribution of 60% which Walker (1975a) ascribes to the chionophilous tendency of certain sedges. Sites with substantial percentages of "other herb" pollen at the *Artemisia* maximum are in several cases the same as those with the highest values for *Empetrum* plus *Salix* and for other shrub and tree pollen, or occur in the same general areas of the Highland margins. Walker (1975a, 1977) has suggested that extensive snow-beds exerted a strong influence on the stadial vegetation at the Blackness and Roineach Mhor sites and Brown (1971), with the evidence then available, used stadial *Empetrum* percentages as evidence for continued oceanicity in NW Scotland. It is highly significant that firn-line calculations suggest substantial stadial snowfall in these same areas (Sissons and Sutherland, 1976; Sissons, 1977a).

North and east of the main watershed, indicated in Fig. 3 by ice-caps and glaciers beyond the main ice mass, values for *Empetrum* plus *Salix*, together with those for other woody taxa and "other herbs" are generally lower than at the S and NW margins of the Highlands. *Artemisia* values, by contrast, are generally higher in the lee of the watershed, reaching over 60% at the Abernethy Forest site. These high values led Birks and Mathewes (1978) to suggest low precipitation in this area of the Highlands; Sissons' suggestion (this volume) of an annual precipitation of only 200-300 mm in Speyside lends support to their view. Thus the pollen data reveal a coherent geographical pattern, the climatic implications of which are consistent with the evidence provided by firn-line calculations. It would seem, therefore, that to a large extent the stadial pollen spectra represent contemporaneous pollen production rather than secondary pollen deposition. It follows that at the extreme of the stadial there was a marked contrast between windward regional slopes with more oceanic conditions, or at least heavy snowfall, and leeward areas with more continental conditions.

ENVIRONMENTAL CHANGES AT THE TOM NA MOINE SITE DURING THE LOCH LOMOND STADIAL

There are problems associated with interpretations of the pollen spectra from Tom na Moine, *viz.* an absence of absolute pollen data to check pollen concentration changes, relatively low pollen sums, and often wide stratigraphic intervals between spectra. Nevertheless there appears to be a meaningful pattern in the spectra from the lowest 3 pollen assemblage zones, and it is argued below that ecological interpretations of this pattern are supported by lithostratigraphic data. The variations outlined below suggest a tentative model that may have wider application to stadial sediments in Scotland.

While the *Artemisia*-maximum spectrum from the Tom na Moine core reveals a continental tendency, at earlier and later phases of the stadial there are higher sums of pollen of chionophilous taxa. This is brought out more clearly in Fig. 4 which provides a synoptic view of the stadial pollen spectra. On the assumption that

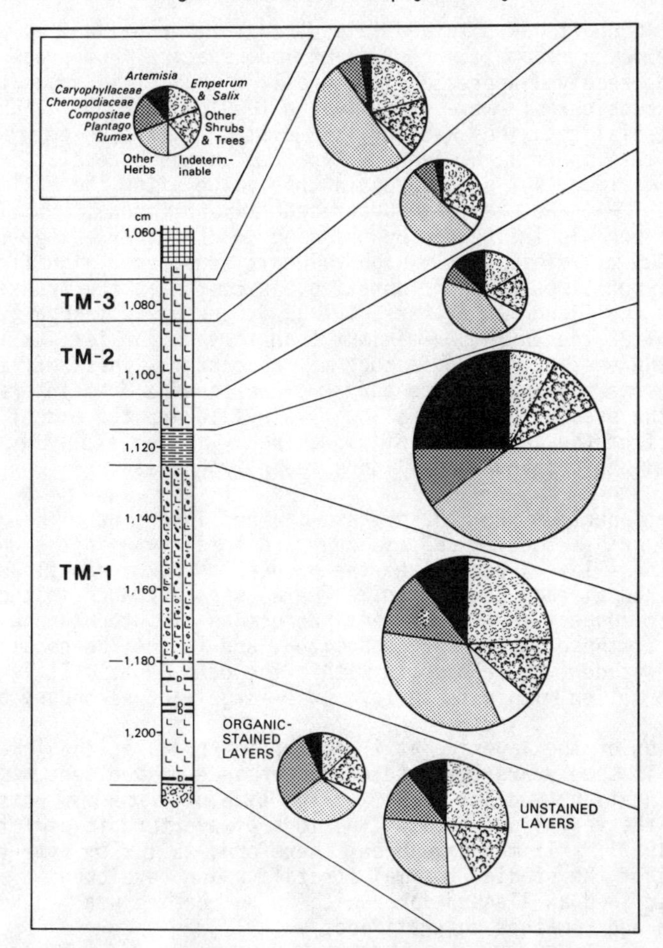

Fig. 4 Mean pollen spectra from the minerogenic sediments at the base of the Tom na Moine core.

the stratigraphic subdivisions within pollen assemblage zone TM-1 are of ecological significance, pollen counts for each subdivision and for the whole of pollen assemblage zones TM-2 and TM-3 were pooled, and mean pollen spectra are shown by divided circles. The symbols within the sectors represent the same pollen groups as in Fig. 3, with the addition of indeterminable pollen. Two circles are shown in the lowest subdivision, representing the mean pollen spectra from the four layers of organic-stained clay, and from the unstained clay. The areas of the circles are proportional to the mean number of land pollen grains per slide; no scale is given, since this is a very crude measure of pollen concentration.

If the mean spectra from the Tom na Moine core are compared with those from the *Artemisia*-maximum horizon throughout the Highlands, it may be seen that it is only the single spectrum from the laminated silty clay between 11.15 and 11.25 m which is markedly continental. It is normally questionable to place much emphasis on a single pollen spectrum, particularly where a pollen sum of only 100 grains is available, but this spectrum is considered significant because of the unusual nature of the sediment from which it was obtained (discussed further below). Earlier and later stadial mean spectra resemble *Artemisia*-maximum spectra from areas which, it has been suggested, received appreciable snowfall. Thus it is suggested that snow-beds may have been considerably more extensive in the Tom na Moine catchment early and late in the stadial than they were at its maximum. Differences are apparent between the early and late spectra, however. *Empetrum* values generally exceed 20% before the *Artemisia* maximum, but do not regain this value after the maximum until the time of pollen zone TM-3, the Gramineae-Cyperaceae-*Empetrum* zone. The mean *Empetrum* value for this zone is influenced by the high *Empetrum* percentage (39%) for the sample at 10.825 m. This single high value is considered significant since an *Empetrum* peak, sometimes of short duration, is common at the transition from the Lateglacial to the Flandrian (Vasari, 1977). Cyperaceae make a greater relative contribution after the *Artemisia* maximum than they do earlier, as is seen by the values for "other herbs". Two factors may account for these differences: there may have been some release of secondary *Empetrum* grains from interstadial soils at the onset of the stadial, while the lower values toward the end of the stadial could have resulted from the inhibition of flowering by longer snow-lie, a circumstance which could well have favoured the increase in Cyperaceae.

The mean pollen spectrum from the organic-stained layers near the base is anomalous; percentages of both *Artemisia* and *Empetrum* are lower than in the unstained clay within which these layers occur, and the proportion of Caryophyllaceae is exceptionally high, as has already been noted. There is no evidence in the curves for aquatic taxa to suggest any significant production of autochthonous organic material at the time that these layers were deposited, and it must be concluded that the organic stain was derived from soils within the catchment. It is possible, therefore, that the pollen within these layers is largely of secondary origin.

The preservation of the layering at irregular intervals in the lake sediment suggests that in some years the rate of disturbance of sediment was less than the rate of sedimentation (Ludlam, 1976). The time of potential disturbance is restricted to the vernal and autumnal periods of overturn in dimictic lakes (Hutchinson, 1957). It may have been, therefore, that from time to time during the early part of the stadial, thermal stratification developed in summer; Hobbie (1961) has described an Alaskan lake which in successive years developed thermal stratification and remained unstratified.

There is no evidence of the preservation of laminations of any sort in the overlying silty clay with grit particles. The nature of the sediment suggests more profound disturbance of the soil than previously, while the relatively high pollen concentration suggests a slow rate of sediment accumulation possibly combined with some influx of secondary pollen. Livingstone *et al.* (1958) have described turbid Alaskan lakes to which sediment is supplied by solifluction processes, by stream action and by localised thawing and slumping of the shores. No thermal stratification develops in summer despite warming to 13°C and the whole water body is subject to wind mixing. Under such conditions it is likely that any fine sediment deposited beneath the winter ice would be resuspended during the following summer, and the sediment would appear unlaminated. It is possible to suggest, therefore, that while the silty clay with grit particles was accumulating summers were cool and perhaps windy, and that winters were sufficiently snowy to protect *Empetrum* and to provide moisture for amorphous solifluction.

By contrast, the overlying silty clay, which contains the *Artemisia*-maximum pollen spectrum, is finely laminated. It is regretted that the laminations were neither measured, counted, nor examined closely, and that only one sample for pollen analysis was taken from the sediment. Laminated sediments of this type usually occur only in basins known to have been fed with glacial meltwater (Pennington, 1977b). Their preservation in the Tom na Moine depression indicates a reduction in the period of free circulation compared with conditions during the deposition of the underlying "solifluction clay"; this could have resulted from a delay in the melting of lake ice until after the period of maximum solar radiation (Hobbie, 1961), in which case the lake is likely to have been of the cold monomictic type. The fineness of the laminations and the absence of coarser particles suggest reduced run-off and solifluction compared with the early part of the stadial, supporting the evidence of continentality provided by the pollen spectrum.

A sharp reduction in pollen concentration (grains per slide index) is indicated at the top of pollen zone TM-1, in the unlaminated silty clay overlying the laminated material (see Fig. 2), and this may be ascribed to a more rapid rate of accumulation of sediment. The pollen spectrum from this sediment is suggestive of a return to the heavier snowfalls of the early part of the stadial, providing increased spring run-off and increased discharge of mineral particles into the basin. The absence of visible laminations in the sediment introduced into the lake at this time may have resulted from a lengthening period of summer circulation, or to shallowing of the water body, or to a combination of both.

There is no visible change in the nature of the sediment in which were deposited the pollen assemblages of pollen assemblage zone TM-2 (Gramineae-Cyperaceae zone) and pollen assemblage zone TM-3 (Gramineae-Cyperaceae-*Empetrum* zone) but the greater pollen concentration in pollen assemblage zone TM-3 may indicate a decrease in the rate of sedimentation, probably combined with an increase in pollen productivity.

The Tom na Moine site is unusual in that minerogenic sedimentation continued until the time of the birch rise in pollen zone TM-4, the *Juniperus* zone. At the other sites in the Spey valley listed in Table 1, for example, there was already an appreciable organic component in sediment accumulating in the later part of the stadial. This difference may be ascribed to the slower development of stable soils on the steep slopes within the Tom na Moine catchment.

Summary of Environmental Changes

The stratigraphic and pollen evidence from the Tom na Moine core suggests the following tentative sequence of environmental phases during the Loch Lomond Stadial:

1. During the early part of the stadial snow-beds appear to have been quite extensive, providing moisture for the transportation by overland flow of fine material and organic soil particles into the lake basis; the lake developed thermal stratification in some summers.

2. Later, soils were disturbed more deeply and coarser particles were included in the lake sediments; the lake appears to have become monomictic, probably because of lower summer temperatures, with the possible added factor of stronger wind disturbance.

3. During the most extreme conditions of the stadial, the *Artemisia* maximum, snow-beds appear to have been reduced; reduced run-off led to a decrease in the rate of sedimentation, and the lake was probably of the cold monomictic type with a short ice-free season.

4. Snow-beds appear to have increased again in the last phase of the stadial, but soil disturbance was less severe than in the earlier phases. There is no evidence of the return of the lake to dimictic conditions, but this may have been due in part to a reduction in its depth.

Discussion

The Tom na Moine site is well placed to have registered environmental changes, since a considerable range of habitats is included in its catchment, and since it lay close to the transition between more continental and more oceanic areas at the climax of the stadial. It is possible that the apparently more continental conditions registered at the stadial climax were related to the culmination of the Younger Dryas readvance of the oceanic polar front (Ruddiman *et al.*, 1977), and that the suggested periods of heavier snowfall which preceded and followed the cold, dry conditions of the climax were associated with the readvance and retreat of the front. Windward regional slopes continued to receive appreciable snowfall even at the stadial climax, however, as is seen by the proportions of the pollen of chionophilous taxa in the *Artemisia*-maximum pollen spectra (Fig. 3).

Sissons (1974b) writes that "after reaching their maximal extent, many glaciers remained active for a time while their margins retreated slightly with intermittent halts, or more probably, minor readvances". His evidence is the presence of more than one end moraine in the terminal zones of a number of Loch Lomond Stadial glaciers (Sissons, 1972, 1974a, 1974b, 1976b, 1977a, 1977b, 1977c; Sissons and Grant, 1972; Sissons and Sutherland, 1976). The two periods of heavier snowfall suggested in this paper would have resulted in two pulses of accumulation, and indicate a possible partial explanation of the suggested glacial response.

ACKNOWLEDGEMENTS

The writer expresses gratitude to the Department of Geography, University of Edinburgh, for making available field equipment and laboratory facilities. The paper owes much to discussion with Dr. M.J.C. Walker, who at an earlier stage of the work, acted as a field assistant and provided advice on pollen identification. Field assistance was also provided by R. Cornish and Dr. A.G. Macpherson; the illustrations were drawn by G. McManus.

Interpretation of the Lateglacial Marine Environment of NW Europe by Means of Foraminiferida

A. R. Lord

(University College London)

ABSTRACT

The Lateglacial is in geological terms very short and in absolute chronology closely defined. Within north-west Europe planktonic foraminifera are rare or absent and study is consequently confined to bottom-dwelling species. As these species are essentially all extant, biostratigraphic studies depend upon the interpretation of sequences in terms of environmental change as indicated by the benthonic foraminifera present. Correlation of such 'environmental stratigraphy' over any distance is unreliable unless underpinned by absolute chronology. The evidence provided by foraminifera for conditions in the marine environment of the Lateglacial North Sea and adjacent waters is reviewed, although it is at present fragmentary. New information from the Shell-Esso South Cormorant Field close to the continental shelf edge is discussed.

INTRODUCTION

During preparatory work for the Quaternary Research Association symposium on Lateglacial environments, two particular aspects became apparent about which information was at best inadequate. First, the number of sites for which useful foraminiferal data are available is small and, secondly, there is no oceanographic or climatic model to provide a context within which to consider the marine microfauna. Following discussion with Dr. J.S.A. Green (Imperial College), and at the symposium, it became clear that construction of a physical model was hampered by lack of biological input. Thus the present paper sets out to discuss the analysis of foraminifera for environmental purposes and to collate what foraminiferal evidence is available for Lateglacial time in north-west Europe.

Charles Lyell, in his classic subdivision of the Tertiary, made use of the fact that the proportion of extant species in a fossil assemblage increases through Cenozoic time as you approach the present day. He was using molluscs but the same trend can be seen in the record of all animals and plants, including foraminifera. In the early Quaternary of the southern North Sea Basin, certain species (*Cibicides lobatulus, Ammonia beccarii, Quinqueloculina seminulum, etc.*) occur which still live in the North Sea today. By contrast, the stratigraphically useful form *Elphidiella hannai* migrated into the area during the Pliocene and is still alive in the northern Pacific today, but disappeared from the North Sea Basin towards the end of the early Quaternary. There is also an element in the fauna of species which are extinct and not known since pre-glacial times (*e.g. Pararotalia serrata, Elphidium haagensis*). For the later Quaternary, the species found are all extant, although frequently living now at higher latitudes or greater depths than in the past, and they therefore cannot be used in the classic appearance/extinction system of biostratigraphy. Instead, it is possible to use knowledge of modern distributions and habitat preferences to study patterns of environmental change as indicated by the fossil foraminifera.

Foraminifera, in conjunction with calcareous nannofossils, have been used in the CLIMAP (Climate Long Range Investigation, Mapping and Predictions) reconstructions of sea-surface temperature at 18,000 years BP (CLIMAP team, 1976; Cline and Hays, 1976). This study is based upon interpretation of assemblages of planktonic foraminifera and coccoliths from levels in cores dated by $CaCO_3$ curves, C14 and $^{18}O/^{16}O$ to 18,000 BP. The species are all still living and it is therefore possible to deduce fairly accurately conditions in upper oceanic waters at the time of the last glacial maximum. Interesting and important as this work is, it is difficult to extrapolate the palaeoclimatic patterns obtained, or the faunal, floral and sedimentary variations, from an oceanic to a shelf setting. CLIMAP is based on oceanic sediment cores which are the product of slow continuous deposition and such a system cannot easily be applied to sediments deposited in less stable shelf environments. The Quaternary North Sea was a largely land-locked shelf sea and in the fossil foraminiferal assemblages, pelagic forms are rare or absent. Occasional specimens or even influxes of planktonic foraminifera occur, *e.g.* in the early Quaternary Crags of East Anglia (Funnel, 1961, Table 4; Beck *et al.*, 1972, p.138) *Globigerina bulloides* and *G. pachyderma* occur indicating brief episodes of increased oceanic influence. By contrast, in the Hoxnian sediments of the Inner Silver Pit, southwestern North Sea (Fisher *et al.*, 1969) and the Nar Valley Clay of Norfolk (Lord and Robinson, 1978) and in the Weichselian sequences of Vendsyssel, north Denmark (Feyling-Hanssen *et al.*, 1971) only benthonic foraminifera were found. Thus, most of the following discussion concerns bottom-dwelling species and the reconstruction of marine bottom environments. It should, however, be noted that none of the few available offshore boreholes have been examined in detail, as oceanic cores have been studied, and detailed information from the outer shelf area would be invaluable.

THE ANALYSIS OF BENTHONIC FORAMINIFERAL ASSEMBLAGES

The investigation of Quaternary benthonic foraminifera and their environmental interpretation has been pioneered by such workers as R.W. Feyling-Hanssen, B.M. Funnell and J.H. van Voorthuysen, and in turn their work has been based on modern distribution data provided by many others. Interpretation of the fossil assemblages depends upon using as reliable data as possible and these can be influenced by a number of factors:

(a) Statistical reliability. A complete assemblage of at least 300-400 individuals is necessary for statistical comparability between samples, and it is desirable to completely pick a random sub-sample. Flotation techniques are unreliable, except as a means of rapid concentration of tests and the heavy residue should always be picked for specimens left behind. In any case, in the final assemblage obtained agglutinated forms are likely to be underrepresented or even absent, as their tests are very susceptible to post-mortem disaggregation and also to destruction during laboratory sample preparation. The convenience of modern foraminifera with hard tests falling into three groups which can be plotted on triangular diagrams (see Murray, 1973, Figs. 7 and 102) can also be applied to mid- and late-Quaternary assemblages, but in all cases the agglutinated forms (Textulariina) will be less common in the thanatocoenosis than in the original biocoenosis.

(b) Reworking. Reworking of tests, whether at time of death or later is a constant problem and assemblages must be inspected in terms of preservation, wear, colour, population structure, *etc.* to avoid misinterpretation of mixed samples; often misinterpretation may be unavoidable.

(c) Sampling. Sample density can radically affect appreciation of temporal variation in faunas. Isolated samples are of little value compared to a sample sequence documenting a pattern of change in bottom conditions at one site.

(d) Taxonomic Comparability. Comparative studies depend upon the use of other workers' data. In this context, taxonomic uniformity is vital and has been described by Neale (1964, p.258) as a 'Fundamental Factor' in ecological and distributional studies. It is difficult to over-emphasise the importance of accurate taxonomic work. Most of the foraminifera found in the Quaternary of north-west Europe are well known, but not all are easy to interpret and, for example, confusion over the specific differentiation of *Elphidium* types and related forms continues to be a problem. *Elphidium clavatum* is a commonly-cited species with cool water, arctic/subarctic connotations, and is equally commonly misidentified. Unillustrated lists of species are difficult to evaluate and use.

Thus it is possible to build up a pattern of environmental change if a sequence of foraminiferal assemblages is available from a site. Such patterns of change, an environmental stratigraphy based on foraminifera, can be extrapolated and correlated over limited distances. However, local factors can seriously affect the geographic extent of a recognised effect, just as sample density controls the refinement achieveable of patterns of change through time. Such stratigraphies can be seriously misleading. For example, a warming or cooling trend in bottom waters indicated by fossils could occur in, or within, one of several glacial or interglacial episodes. The temptation to place a time significance on, say, a warming trend must be set against the stratigraphic and biological background of the site, and where possible confirmed by absolute dating (*cf.* Morrison, 1969, p.370 on C14 dating and the Pleistocene/Flandrian boundary). Thus, in the next section particular attention is paid to Lateglacial localities with foraminifera for which absolute dates are available. Other sites provide useful comparative information, but are much less valuable for accurately reconstructing conditions over a wide area.

LATEGLACIAL MARINE SITES

The Lateglacial is in geological terms a very precise time-span, the transition from the cold climate of the Last (Devensian/Weichselian) Glaciation to temperate conditions, conventionally about 13,000 to 10,000 years BP. Obviously such a 'time' interval, dependent for its definition upon a climatic effect, will vary in absolute chronology from area to area and therefore the use of C14 dates provides a means of underpinning a series of comparative analyses which are to be combined to reconstruct a pattern of events over a wide area.

In the interpretation which follows, it is important to note that Feyling-Hanssen and colleages use the term 'diversity' as defined by Walton (1964), *i.e.* the number of species which collectively compose 95% of an assemblage. Diversity here is used in a less formal sense and refers to the total number of species present, excluding obvious reworked material.

Sites with Lateglacial foraminifera from the circum-North Sea area (Fig. 1) fall into the following groups:

(a) Sites lacking C14 dates, mostly North Sea boreholes, for which foraminiferal data is available giving some evidence about general trends through the Lateglacial interval - British Petroleum (Forties Field) DB6, Shell-Esso (Auk

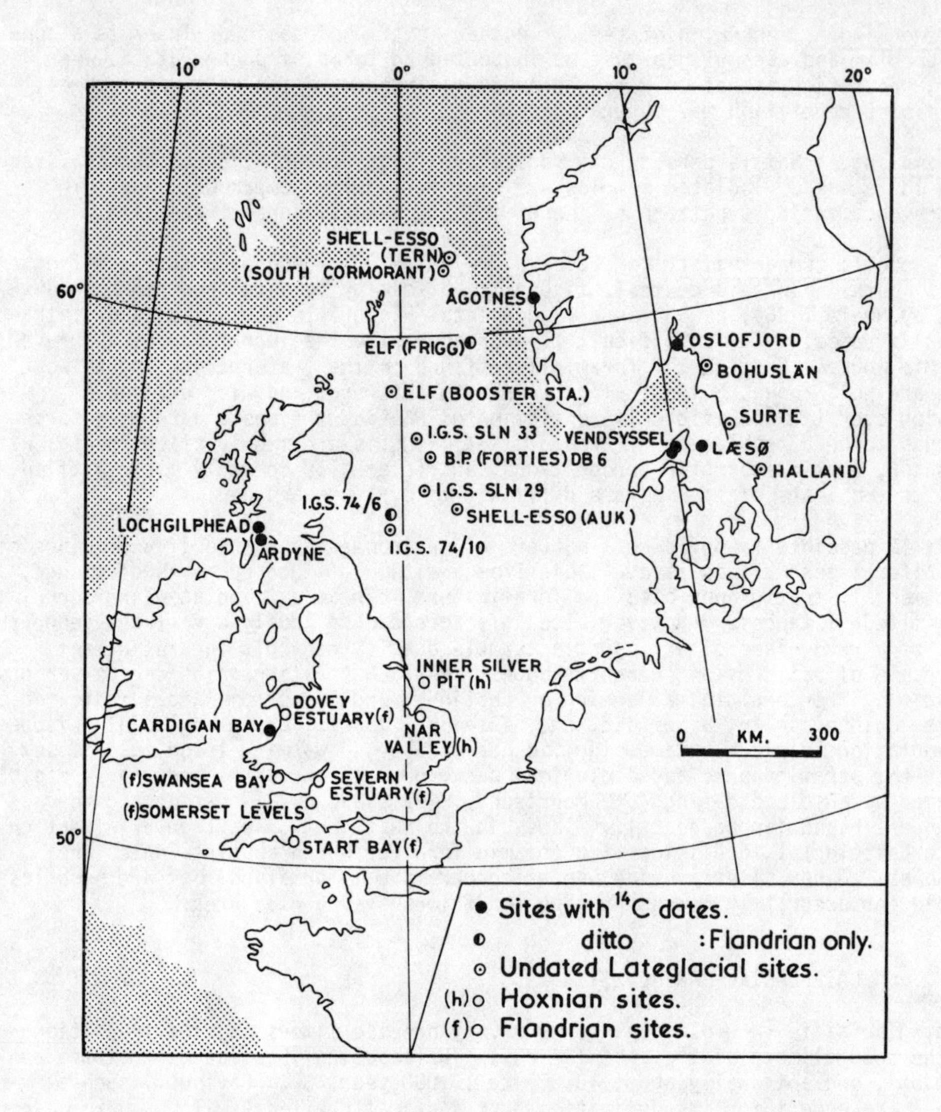

Fig. 1 Lateglacial sites in north-west Europe

Field) 30/16-25, Institute of Geological Sciences (IGS) SLN29 and 74/10 (all reported in Hughes *et al.*, 1977), IGS SLN33 (Harland *et al.*, 1978; Gregory and Harland, 1978), Elf (Booster Station) (Løfaldli, 1973) and onshore sites in south-western Sweden (Hessland, 1943; Brotzen, 1951). New undated information from the Shell-Esso South Cormorant Field is given below.

(b) Sites for which C14 dates are available but only for Flandrian horizons - Cardigan Bay (Haynes *et al.*, 1977); IGS 74/6 (dated level 7,109 ± 60 BP, Hughes *et al.*, 1977) and Elf (Frigg Field) (dated level 7,020 ± 100 BP, Løfaldli, 1973, p.25 - date considered young). Information value as (a) sites.

(c) Dated sites, discussed in detail below - Lochgilphead; Ardyne; Ågotnes, western Norway; Oslofjord, southern Norway; Vendsyssel, northern Jutland

and Laesø, Kattegat, Denmark. Lithostratigraphic sections for these localities, based around C14 dates, are given in Fig. 2.

Cardigan Bay. Haynes *et al.* (1977) examined foraminifera from three offshore sites in Cardigan Bay (IGS 73/42 and IGS cores ZZ27 and ZY23). The site closest inshore (ZZ27) contained a peat dated at 8,740 ± 110 BP, but the first foraminifera were recovered from above the peat. By extrapolation of the dated level to another site (73/42) a little further offshore, a sequence of foraminifera-bearing sands and silts resting on Devensian till can be related to the latter part of the Lateglacial. The sands contain an essentially temperate, shallow water assemblage in which open marine species occur with brackish water forms, while in the silts above, the brackish water element declines and general diversity increases. The silts contain assemblages closely comparable to those found in shallow-water and on muddy tidal flats in Cardigan Bay today. Planktonic forms occur; *Globigerina pachyderma* and *G.* sp. are not uncommon and *Globigerinita bradyi* also occurs. In fact, this is a sequence representing the Flandrian transgression, with evidence for commencement in this area in the last part of the Devensian. This inference rests upon the not very satisfactory correlation of the dated peat, and in this area the Flandrian transgression may have actually removed Lateglacial sediments. Other sites studied around south-west Britain, and which have yielded foraminifera, further document this transgressive episode - Start Bay, Devon (Lees, 1975), Severn Estuary (Murray and Hawkins, 1976), Somerset Levels (Macfadyen, 1955), Swansea Bay (Macfadyen, 1942; Culver and Banner, 1978) and the Dovey Estuary (Adams and Haynes, 1965; Haynes and Dobson, 1969), and provide valuable background information for Lateglacial faunas.

Western Scotland - Lochgilphead and Ardyne. At Lochgilphead a well-dated sequence of 'Clyde Beds' contains a good invertebrate fauna, including foraminifera reported by Wilkinson (*in* Peacock *et al.*, 1977). The foraminifera are dominated by *Elphidium* and *Protelphidium* species, with subordinate miliolids and, in the lower part, *Cibicides lobatulus*; diversity is also greatest in the lower portion of the sequence. Marine bottom temperatures were lower than those at the present day in this area. A shallowing and warming trend can be recognised in the higher part (*c.* 12,000 to at least 11,000 BP) culminating in a foraminiferal fauna in the highest sample containing *Ammonia batavus (= A. beccarii), Miliammina fusca, Protelphidium anglicum* and *Elphidium asklundi*. This site was close to shore and doubtless greatly influenced by local factors.

Another well-dated site at Ardyne, 30 km south-east of Lochgilphead, exposed sediments laterally equivalent to the 'Clyde Beds' and also contained good foraminiferal assemblages (Wilkinson *in* Peacock *et al.*, 1978). The assemblages are generally more diverse than those at Lochgilphead, but are again dominated by *Elphidium* species *(E. clavatum, E. bartletti, E. macellum, E. subarcticum)* and *Protelphidium orbiculare*, with miliolids and agglutinated forms as minor constituents. This locality is stratigraphically complex, with non-sequences and possibly condensed levels, but five units are recognised. The lowest unit (1) contains a sparse fauna dominated by *E. clavatum* and *Cassidulina* which becomes slightly more diverse higher up. The top of Unit 1 is dated 12,171 ± 53 BP. Unit 2, the mid-part of which is dated at 11,575 ± 102 BP, contains good assemblages dominated by *E. clavatum* and *P. orbiculare* indicative of cool, shallow water, but in the upper part of the unit *A. batavus* (= *A. beccarii*) and *E. macellum* appear, suggesting more temperate conditions. Unit 3, dated 11,159 ± 47 and 10,801 ± 67 BP, is poorly known although one sample is dominated by *A. batavus* which gives a distinctly temperate aspect to at least part of the unit. In contrast the foraminifera from Unit 4 indicate shallow, cold bottom waters with arctic and subarctic forms present. The upper part of Unit 4 is dated from 10,412 ± 136 to 10,195 ± 117 BP. Unit 5 (Flandrian) contains a shallow-water, temperate marine fauna. Bottom temperatures were generally much lower than in the sea in the area today. Peacock *et al.* (1978, p.23 and Fig. 5) link the temperate assemblage from Unit 3, about 11,000 BP, with the final part of the

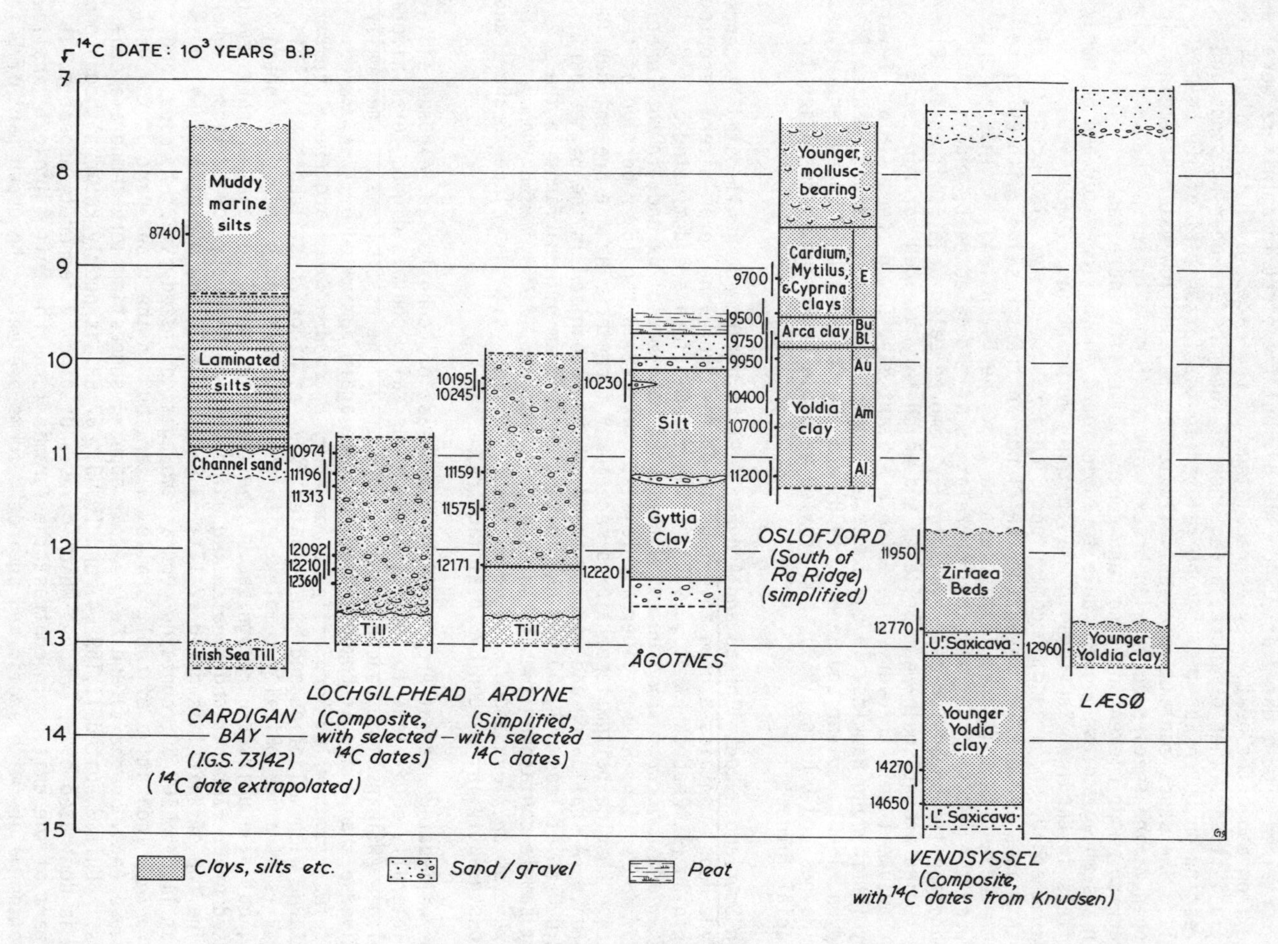

Fig. 2 Dated sites containing foraminifera. The ^{14}C time scale refers only to the dated levels.

Lateglacial Interstadial. Planktonic forams were not reported from Lochgilphead or Ardyne.

Ågotnes, western Norway. Dated marine sediments containing foraminifera from the western coast of Norway have been studied in detail by Mangerud (1977). A gyttja clay (which includes a shell lens dated 12,220 ± 150 BP (Alleröd)) contained an abundant, but not very diverse fauna indicative of cool, but not cold, shallow water, and a sand and gravel lens above had similar assemblages. In contrast, a sample from a shell lens dated 10,230 ± 180 BP (Younger Dryas) contained a slightly more diverse assemblage dominated by *Cibicides lobatulus* and *Astrononion gallowayi*, a composition similar to that of assemblages recorded from fjord bottoms close to glaciers in Svalbard (Nagy, 1965, p.113 and Table 1; Mangerud, *op. cit.*, p.41). Mangerud was able to deduce that warm Atlantic waters influenced this locality from before 12,600 BP, but the foraminiferal evidence (only benthonic forms recorded) is equivocal in this respect. Local influences were probably dominant over bottom water conditions.

Oslofjord, southern Norway. Foraminifera from this area have been discussed in detail by Feyling-Hanssen (1964, 1972). For the present study the southern Oslofjord area south of the Ra morainic ridge provides the fullest sequence, with Yoldia Clay the oldest unit exposed subdivided on foraminifera into three subzones, lower, middle and upper (Al, Am, Au). The whole of Zone A is dominated by *E. clavatum* and *Cassidulina crassa*, while the middle subzone is differentiated by means of higher diversity (Feyling-Hanssen, 1964, pp.379-383. Al 18, Am 48 and Au 25 species) and more abundant *Nonion labradoricum, etc;* rare *Globigerina bulloides* also occur in Am. Feyling-Hanssen (1964, pp.152, 170-175) views the poor Al and Au assemblages as arctic in character but influenced by reduced salinity and the concomitant effects of high meltwater input. Zone B follows on from Subzone Au with slowly increasing diversity (Bl 36 and Bu 46 species) in essentially high latitude assemblages. Bottom water conditions were obviously colder than in the modern Oslofjord.

Vendsyssel, northern Jutland, Denmark. The late-Quaternary foraminifera of Vendsyssel have been described in detail (Feyling-Hanssen *et al.*, 1971) in a companion study to the Oslofjord work discussed above. However, the availability of more C14 dates (Knudsen, 1978) necessitates a small revision of the time scale given in the original paper (*op cit.*, Figure 43), and these new dates are incorporated in Fig. 2. At Nørre Lyngby, Knudsen (*op cit.*) differentiates between the Lower Saxicava Sand (E. subarcticum Zone) and the Younger Yoldia Clay (E. excavatum Zone), with foraminifera indicative of fully arctic marine bottom conditions, and the Upper Saxicava Sand (E. albiumbilicatum Zone), with evidence for more moderate environmental conditions which might correlate with the Bölling interstadial. Assemblages from Younger Yoldia Clay at Nørre Lyngby and Gølstrup (Jørgensen *in* Feyling-Hanssen *et al.*, 1971, pp.124-127) and at Løkken (Knudsen *in* Feyling-Hanssen *et al.*, p.132) are dominated by *E. clavatum* (up to 97%) with *Cassidulina crassa*. Diversity is generally low and the samples closely resemble in their environmental implications Subzones Al and Au of Oslofjord (Knudsen, *op. cit.*, p.152). As Knudsen very properly comments, no age correlation can be derived from this faunal similarity. The younger Zirfaea Layers, sampled at Skeen Møllebaek by Jørgensen (*op. cit.*, pp.127-128), contain foraminiferal assemblages still dominated by *E. clavatum* (75-80%) but with greater diversity (12-24 species, with a total of 41 species from 6 samples) and larger numbers of *Cassidulina crassa, E. subarcticum, E. albiumbilicatum* and *E. asklundi*, all indicative of less extreme conditions and complementing the evidence from the Upper Saxicava Sand at Nørre Lyngby. The youngest part of the Lateglacial and the Weichselian/Flandrian boundary fall within a non-sequence and Flandrian Littorina Deposits rest transgressively upon older sediments.

Laesø, Kattegat, Denmark. A series of boreholes penetrating Yoldia Clay and Flandrian sands on the island of Laesø have been studied by Michelsen (1967). A

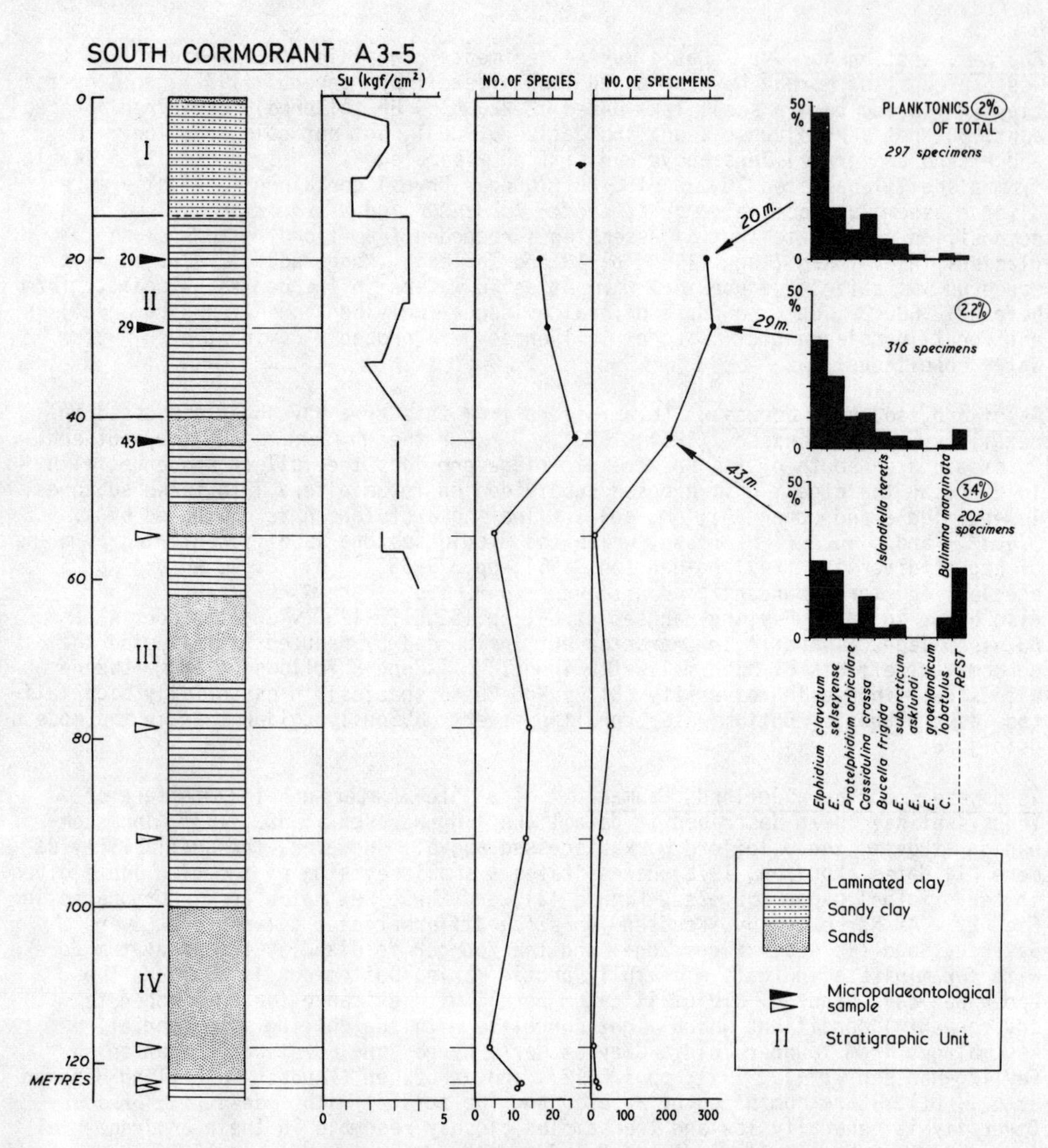

Fig. 3a. Foraminiferal assemblages and undrained shear strength of late Quaternary sediments in the Shell-Esso South Cormorant Field; A3-5 borehole.

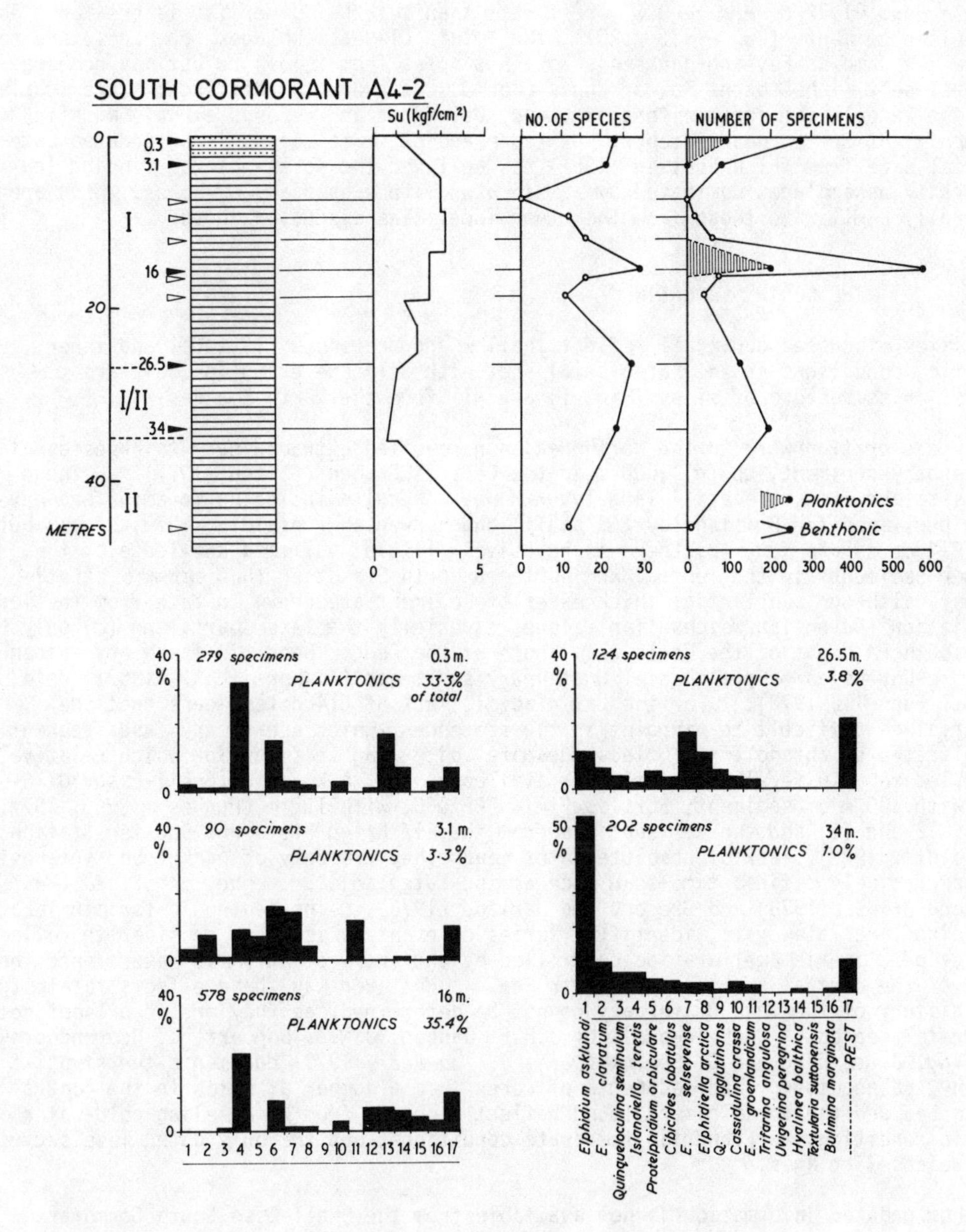

Fig. 3b Foraminiferal assemblages and undrained shear strength of late Quaternary sediments in the Shell-Esso South Cormorant Field; A4-2 borehole.

C14 date for the base of one borehole gave 10,960 ± 180 BC (= 12,910 ± 180 BP). In the assemblages from the Yoldia Clay (*op. cit.*, pl.8) only two contain over 50% *E. clavatum* and in only one case, the deepest sample, does the assemblage closely resemble those from the Younger Yoldia Clay of Vendsyssel, *i.e.* Borehole 10, sample 20: *E. clavatum* 91.7%, *C. crassa* 3.3%, rest less than 10%, but diversity is greater. As Michelsen comments (*op. cit.*, p.251), the Yoldia Clay assemblages from Laesø are not quite the same as Feyling-Hanssen's Zone A samples from Oslofjord but may compare more closely with Subzone Bu, or Zones C or D. However, the differences are subtle and the Lateglacial foraminifera of Laesø, Oslofjord and Vendsyssel have much in common. At Surte, near Göteborg in Sweden and close to Laesø, is an undated Lateglacial site from which Brotzen (1951) has analysed the foraminifera, finding low diversity assemblages dominated by *E. clavatum* with subsidiary *C. crassa* which are generally similar to those from the other localities discussed above.

THE NORTH SEA BASIN

The sites discussed above all reflect shallow and nearshore (littoral and inner neritic) conditions in the Lateglacial and, with only one exception, all are onshore. The residue of sites (Fig. 1) are all from the North Sea Basin.

Early exploration work in the North Sea soon revealed extraordinary thicknesses of Quaternary sediment, up to 1,000 m in the Central Graben (Caston, 1977). These deposits, however, are still largely unknown. Results including foraminifera have been published for British (by IGS staff) and Norwegian (Løfaldli, 1973) areas, but (a) C14 dates are few, (b) there is relatively little published knowledge of the deeper sediments in the central and northern North Sea other than seismic stratigraphy, although substantial thicknesses of sediment are known to date from the last glaciation (Devensian/Weichselian) alone, especially the later part, and (c) only in the southern bight of the North Sea, south of the Dogger Bank, is there any evidence for pre-Devensian or pre-glacial Quaternary sediments (Fisher *et al.*, 1969; Oele, 1971; Funnell, 1972). For the Lateglacial, lack of C14 dates means that the interval is difficult to pinpoint in the sequences, which are in any case frequently complicated by channelling. Thus, despite quite long sections for which relatively detailed foraminiferal information is available, *e.g.* Auk Field (Shell-Esso 30/16-25) with 100 m of sediment, Forties Field (BP DB6) with 130 m (Hughes *et al.*, 1977, Figs. 12 and 13) and shorter sections from the Elf Frigg Field and Booster Station (Løfaldli, 1973), lack of absolute dates means they are only of background interest for a precisely defined time-span such as the Lateglacial. Hughes *et al.* (*op. cit.*), Harland *et al.* (1978) and Gregory and Harland (1978), using benthonic foraminifera and dinoflagellate cysts, identify a series of events which they consider to reflect phases of climatic amelioration controlled by the entry of Atlantic waters into the area of the central and northern North Sea. Whatever way these effects relate to the history of the late Quaternary cannot be determined, as they are at present not accurately dated. Work in-press by J.H.F. Jansen, J.W.C. Doppert, K. Hoogendoorn-Toering, J. de Jong and G. Spaink (*Neth. J. Sea Res.*, 1979) documents the stratigraphy, palaeontology and sediments of cores from a number of sites in the central North Sea and the foraminifera found reflect a change from Weichselian cold-water arctic conditions to Flandrian temperate conditions, but the only dated levels give mid-Weichselian ages.

Further undated information is now available from the Shell-Esso South Cormorant Field (Fig. 3a and b), for despite lack of dates, the site is of great interest as it is close to the continental shelf edge. The two boreholes figured (A3-5 and A4-2) complement each other, in that the latter contains much fuller information for the top part of the sequence. The stratigraphic units are those of Shell staff and are based on seismic reflections and lithological characteristics. In A4-2 two phases of high diversity (16 and 0.3 m below surface) are characterised by specimen abundance and unusually large numbers of planktonic foraminifera (*Globigerina bulloides*,

G. pachyderma with rare *G. quinqueloba*), presumably carried into the area from the Atlantic. The increase at the top reflects Flandrian temperate conditions, although the precise age of the sample might be relatively modern. Associated with the two plankton-rich levels in A4-2 is the appearance of the deep-water benthonic species *Uvigerina peregrina*. This species is known today essentially from the outer shelf and shelf edge in waters up to 1,000 m depth, and is important for oxygen isotope analyses (Shackleton, 1977, p.5). In the South Cormorant Field (Fig. 3a), Units I-III consist of silty clays, laminated clays and fine sands resting on fine to coarse sands with pebbles and shells of Unit IV. The assemblages obtained from III and IV are poor, which is especially disappointing for IV, which is a substantial sand at least 30 m thick. In the nearby Shell-Esso Tern Field (Fig. 1) borehole A1-1 penetrated a sequence of clays resting on an important sand containing the following assemblage at 121.5 mbs: *Islandiella* spp. 55%, *Elphidium subarcticum* 7%, *E. selseyense* 6%, *Cibicides lobatulus* 5%, *Trifarina fluens* 4%, *Buccella frigida* 4%, *Cassidulina crassa* 3%, *E. clavatum* 3%, *Quinqueloculina seminulum* 2%, Planktonic forms 10% (247 specimens; Benthonic spp. 24, Planktonic spp. 2). The clays above contain poor assemblages of cold water aspect. The relatively high diversity and number of foraminifera in the sand at 121.5 mbs, together with some temperate elements, make it tempting to assign an Ipswichian/Eemian age, but there is no direct evidence. The assemblages depicted in Fig. 3b) for Units I and II in South Cormorant A4-2 must, in some cases, be coeval with those described from Ågotnes, Oslofjord, Vendsyssel and Laesø, yet are very different in aspect, being more diverse and not overwhelmed by *Elphidium clavatum* and *Cassidulina crassa*. This must reflect more open marine, mid- to outer-shelf conditions. Whether the pulses of planktonics and associated *U. peregrina* reflect deepening or increased current-dependent, oceanic influence is not clear, although the presence of *Hyalinea balthica* (which is used to mark the base of the Quaternary in the Mediterranean) and absence of *Protelphidium orbiculare* suggest warmer conditions and not deepening. The South Cormorant sediments are channelled between Units I and II. Channelling, so common in other parts of the North Sea (Holmes, 1977, Fig. 5; McCave *et al.*, 1977. p.194, Figs. 14.4 and 14.5) may occur at several levels with different flow directions at one site. It is generally attributed to subglacial water cutting through soft sediment, in the so-called 'tunnel valley' style, as it seeks an outlet from beneath ice. The sediments also show marked variation in undrained shear strength (Fig. 3a and b; Shell data). The fluctuations in shear strength indicate that at certain levels the sediment is overconsolidated, *i.e.* the shear strength of the sediment is greater than would be expected for the overburden present (see Greensmith and Tucker, 1971, pp.743-44), which may be due to weathering and erosion, but in the present case could be caused by ice-loading. The foraminifera were examined to see if they could provide evidence of marine conditions which might support or refute the idea of ice-loading being responsible for overconsolidation of the sediments. In practice, the samples were taken at far too large intervals for the environmental significance of individual samples to be combined into a useful pattern. It was not therefore possible in the present case to correlate cold water conditions, as indicated by the foraminifera, with zones of overconsolidation of the sediments. The sediments themselves are not directly helpful. The fine-grained sediments appear to be normal marine clays, generally lack 'stones' and have microfaunas compatible with marine, if cold water deposition. Were the clays laid down beneath or in front of ice and is it possible to distinguish between two such clay types? The sedimentary and geotechnical properties of onshore tills cannot be applied to marine deposited glacio-marine clays, which can make the sequences difficult and confusing to interpret.

CONCLUDING REMARKS

At present, information from the North Sea Basin has complicated as much as extended our knowledge of the Quaternary of north-west Europe. Direct comparison of the deep sea record with the terrestrial record by means of palynology and isotopic analysis is proving easier than comparing either with shelf data. This is not entirely surprising in view of the particular characteristics of shelf environments. The interpretation of glacio-marine sediments is still uncertain, however, and the significance of the shear strength patterns and of the channelling may be learned from detailed study of sediments and microfossils in dated borehole sequences. Evidence from the South Cormorant Field suggests the importance of borehole material from the outer continental shelf and shelf edge to help link shelf and deep ocean events.

The foraminiferal evidence for the Lateglacial cannot yet be synthesised into a meaningful pattern, as the sites are scattered and most of the dated ones reflect local as much as regional effects. Little is known about how long climatic change, as indicated by terrestrial pollen, takes to affect marine surface or, more particularly, bottom waters and how rapidly benthonic organisms react to, for example, sudden fluctuations in temperature and salinity caused by meltwater input. The information given by Nagy (1965) of foraminifera living within a few metres of, presumably, melting ice in Vestspitsbergen is not necessarily a valid model to extend to the North Sea Basin, which is largely enclosed and was almost surrounded in the Late Devensian/Weichselian by ice-covered land to contribute cold, fresh meltwater. However rapidly mixing of the water column took place, strong variation in marine bottom conditions (salinity, temperature, *etc.*) must have occurred, to the detriment of benthonic organisms. The limited access of planktonic organisms was controlled by the presence and distribution pattern of ice, as well as by variation in strength and direction of oceanic currents, water mass movements and oceanic front fluctuations in the North Atlantic.

ACKNOWLEDGEMENTS

I am grateful to Dr. R.W. Feyling-Hanssen (Århus), Professor B.M. Funnel (East Anglia), Dr. J.R. Haynes (Aberystwyth) and Lektor Karen Luise Knudsen (Århus) for past advice, discussion and help. Miss Fiona Hitchen kindly prepared the manuscript and Mr. C.F. Stuart drew the diagrams. Permission to publish information for the South Cormorant and Tern Fields from Shell UK Exploration and Production Ltd. and Esso Exploration and Production UK Inc. is gratefully acknowledged.

The Marine Ostracod Record from the Lateglacial Period in Britain and NW Europe: A Review

J. E. Robinson

(University College London)

ABSTRACT

For over a century ostracods have been mentioned in accounts of glacial and Lateglacial successions. The history of such ideas and the assumptions behind them are reviewed and summarised. At best, the ostracod record for the Lateglacial offers a climatostratigraphy recording climatic amelioration from the Late Devensian to the Flandrian, with a brief reversal corresponding to the Loch Lomond Stadial. Such complete sequences are rare onshore, but available offshore near the west coast of Scotland. There are still problems when correlation is attempted from west to east coast sequences in Scotland, and correlation is sought with the published North Sea records. Such problems require fuller, more detailed study of continuous cores for ostracod data and the support of absolute age determinations. Ostracod faunal studies in Scandinavia need to be undertaken to provide the information on which to base comparisons for north-west Europe.

In recent years, we have become familiar with the impressive precision in correlation of the Pleistocene of the deep ocean basins achieved by the micropalaeontologists working upon phytoplankton, coccoliths and planktonic foraminifera from ocean-floor cores. Such work is contained in most volumes of results from the Deep Sea Drilling Program (*e.g.* Miles, 1977), but most fully in the CLIMAP Memoir of the Geological Society of America (Cline and Hays, 1976). Naturally, there has been some expectation on the part of Pleistocene workers that similar studies on the Continental Shelf could help answer the existing problem of British-European correlation. The object of this contribution is to explain why application so effective in the Atlantic, does not similarly apply in the North Sea. Localities referred to in this paper are indicated on Fig. 1.

History of Ostracod Studies, and Ideas on Ostracod Ecology

First, it needs to be stated that, in contrast to the planktonic foraminifera, ostracods make up only a very small percentage of deep-water fauna (Benson, 1971), and so contribute only to a limited extent to deep-sea core correlations. Secondly, for the Continental Shelf and for the North Sea Basin in particular, micropalaeontologists have to deal with an ostracod fauna which is 99.9% benthonic, and in its distribution is influenced by ecological controls quite different from those of the open ocean. Between the two realms, there is a sharp dividing line at the edge of

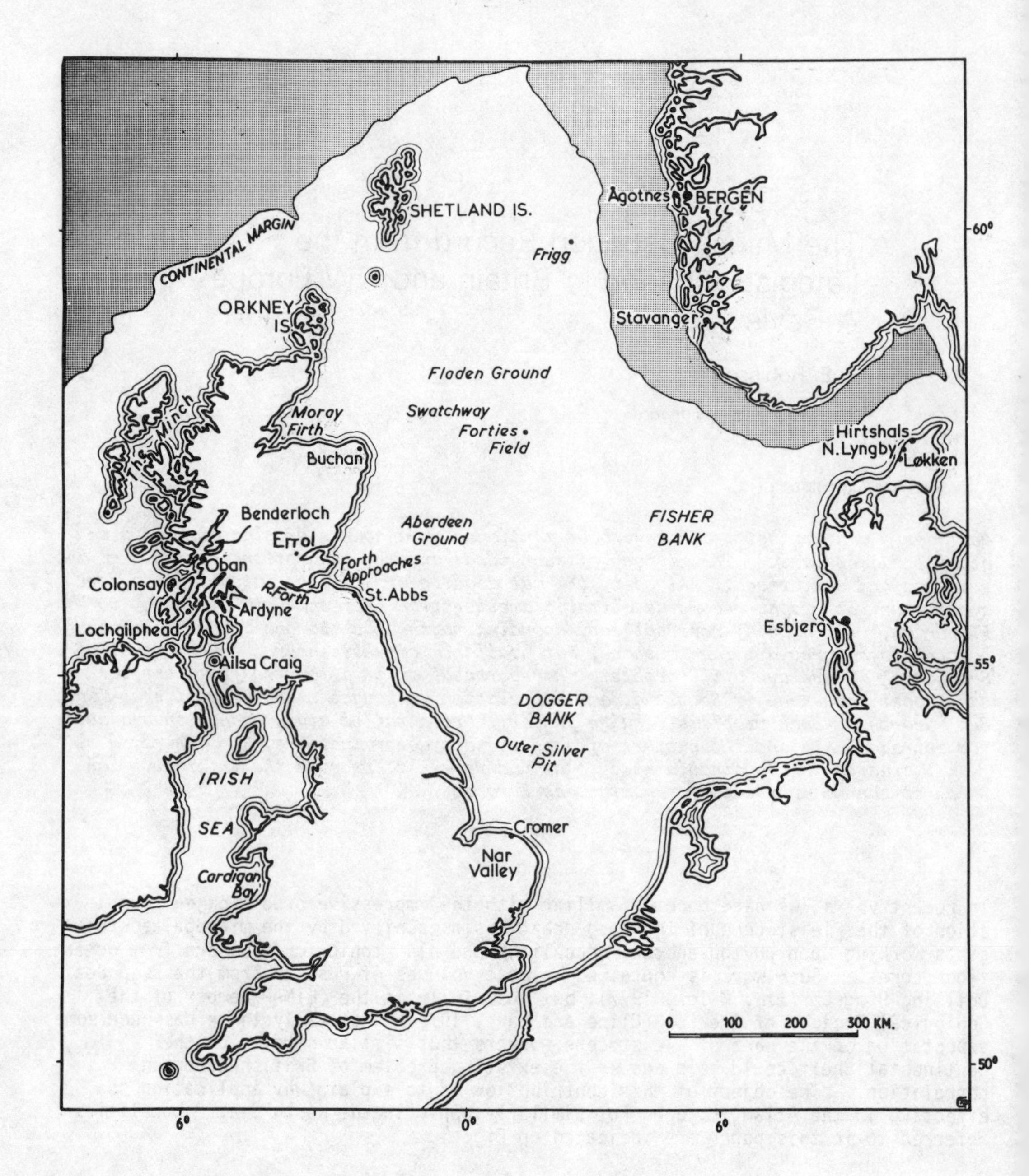

Fig. 1 Location of sites referred to in text

the Shelf, with much less of an overspill than is true for the planktonic foraminifera (Lord, this volume). Thus direct correlation between the Atlantic and the Continental Shelf around Britain is not possible, and it remains to be seen what can be made of the benthonic assemblages of the Shelf in space and time distributions.

The distribution patterns for living ostracod communities are governed by a wide range of factors, some physical, including water temperature, salinities, and nature of the substrate (Neale, 1964), others biological, including food chains, and natural associations with other organisms (Whatley, 1976). As for any ecology, it is difficult to cite any one control as dominant, although water temperature is probably that most widely referred to by ostracod workers. This allows it to be said that marine ostracods on the Shelf do reflect the broadly climatic zoogeographical provinces which were so astutely recognised by Forbes in the last century (Forbes, 1846). Forbes it was who introduced the names for such provinces - 'Arctic', 'Boreal', 'Celtic', 'Lusitanian' - which emphasize the broad correlation between fauna, latitude and ocean water temperature.

From the point of view of our interest in the Lateglacial, it is necessary to add to Forbes' spatial provinces, the understanding of faunal change through vertical successions - the time dimension - which came through the combined work of G.S. Brady, H.W. Crosskey, and D. Robertson in their Palaeontographical Society monograph (1874). Here, they demonstrated that the oldest fossil ostracod assemblages from the Clyde Beds of the Glasgow area were quite different in character from those of present day for the same sea area. As the fossil fauna coincided with some of the "Arctic Shell Beds" of James Smith (1839), in their conclusions it became the first microfaunal evidence for arctic sea water temperatures around the coasts of Britain - a view amply confirmed by later evidence. Three of their most consistent arctic indicator species are *Krithe glacialis* (Brady, Crosskey & Robertson), *Cytheropteron montrosiensis* (Brady, Crosskey & Robertson) and *Rabilimis mirabilis* (Brady), which are found alive and breeding today no further south than the waters of the Barents Sea, or the fjords of East Greenland north of 76° (Hazel, 1967, 1970).

Collectively, the three species may make up the total fauna of sediments deposited immediately following deglaciation in water with temperatures close to freezing and turbid with suspended sediment. The best known example of both the fauna and lithology of this type is the Errol Clay of Tayside, undated, but generally assumed to be about or possibly older than 13,000 BP (McCave *et al.*, 1977; Eden, *et al.*, 1978), but elements of the same association are also to be found in parts of the succession from the Loch Lomond and earlier stadia, as well as older interglacials.

Within any full thickness of Clyde Beds, between the extremes of arctic and present-day temperate conditions, there commonly occur a number of natural associations of ostracods which must represent intermediate climatic regimes for which Forbes' term Boreal (variously subdivided) might be employed. Such fossil faunas would find their living equivalents in the waters off the central and northern coasts of Norway. In a vertical succession, the substitution of one faunal association for another can be interpreted as the replacement of one zoogeographical province by another, much as is demonstrated in the CLIMAP Project for much longer intervals of time equated with stadia or interglacial periods (McIntyre *et al.*, 1976). This history of ideas, and the transfer of what is known of present-day ecology back in time to interpret palaeoecology, is dwelt upon here for the simple reason that the main contribution of ostracod fauna to the understanding of the Lateglacial must be its documentation of a climatostratigraphy for deglaciation. Over short distances, this may be enough in itself to bolster up correlations by standard methods; over wider areas, it may not be totally adequate without accurate C14 dates and other support. What follows is a brief review of the evidence in some key areas.

West Coast of Scotland

In any long sequence of Clyde Beds available from the west coast offshore, it is usually possible to identify the Loch Lomond Stadial temporarily reversing the overall trend of climatic amelioration following the end of the Late Devensian glaciation. Although lacking the affirmation of precise dates, the Clyde Beds sequences of Lochgilphead (Peacock *et al.*, 1977), Inchinnan, and several localities further north in Benderloch, north of Oban (Robinson, unpub.), have a faunal character which allow them to be fitted within a framework of climatic change leading to the Loch Lomond Stadial. Admittedly, such onshore sections tend to be short, and may be broken by unconformities difficult to detect except in exceptionally good exposure such as was available at the Ardyne rig-building site (Peacock *et al.*, 1978), but in some of the Pleistocene sediment traps off the west coast, longer, unbroken sequences are retained, as recorded in several IGS Boreholes (Chesher *et al.*, 1972). In one of these, off Colonsay (IGS 71/9 - See Fig. 2), the cold climate reversal of the Loch Lomond Stadial came between the depths of 16 and 26 m in the cores, marked by a strong fall in ostracod species number as much as by the presence of arctic-indicator species such as *Krithe glacialis* and *Cytheropteron dimlingtonensis*, and included dropstones. Below 26 m, the ostracod species diversity increases to that normal for the Clyde Beds, and the fauna is of species which could be termed Boreal. As yet, the character of the lower clays close to contact with the upper surface of the till locally has not been tested, but should they prove to be arctic by nature, or akin to the Errol Clay discussed above, then this sequence could well serve as a climatostratigraphical standard for the Lateglacial of the west coast of Scotland.

East Coast of Scotland and North Sea

For the east coast of Scotland, it has to be said that there are differences in the Lateglacial record as preserved compared with that outlined for the west. Such contrasts could reflect the absence at an early period of North Atlantic Drift currents in this quarter of the North Sea, currents which, as shown by Laevastu (1963), apart from being a major control in plankton transfer from the Atlantic, also produce temperature contrasts within the North Sea Basin. Otherwise, the heavy flow of clastics from the meltwaters of the eastern Highlands discharged into an essentially shallow shelf sea and produced an east coast succession in which the predominance of sands and silts over clays was certain to govern a different type of benthic ostracod association. The contrast in ostracod fauna was well known to David Robertson, and in several of his accounts of Scottish fauna, he sought to rationalise from established biologic-ecological factors why on the one hand there was an absence of true Clyde Beds on the east coast, just as there was equally an absence of Errol Clay on the west coast (Robertson, 1875). The logical alternative to the speculations by Robertson as to the causes of the contrasts between west and east coast faunas, would be to recognise that they could be of different ages. Currently, this view has some strength in that it is widely held that the Errol Clay predates 13,000 BP (Peacock, 1975a), whereas dates for the Clyde Beds of the west coast extend back to 13,150 BP only on the basis of possibly reworked shells (Browne *et al.*, 1977), the bulk of the clays being no older than 12,750 - 12,400 BP (Peacock *et al.*, 1978). As yet, no exposure or borehole sequence on the west coast has shown typical Clyde Beds underlain by lithologies or fauna which would be typical of the Errol Clay of the east coast. It is in situations such as this that we could benefit from several reliable shell- or wood-dates in order to demonstrate contemporaneity or age difference in successions.

For the east coast, however, there exists at the base of sequences in several areas, the recognisable and well-defined correlation horizon of the Errol Clay immediately overlying till, containing ice-rafted dropstones as well as arctic-indicator ostracod species. What follows higher in the succession seems to represent a great delta complex of clastics in the area of the Tay Estuary and offshore, punctuated by silt and clay horizons, which, while predominantly temperate in their character,

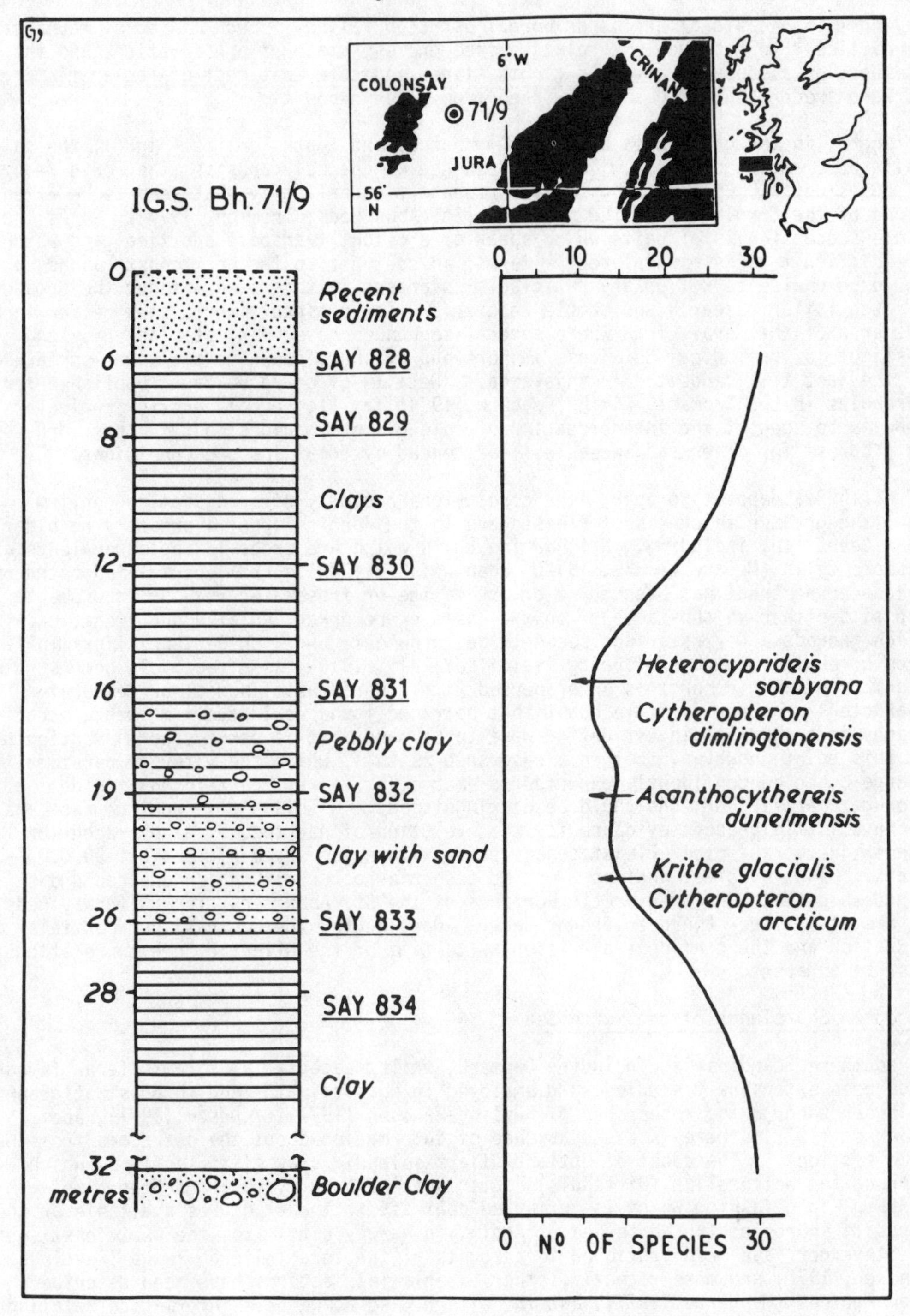

Fig. 2 Lateglacial ostracod species variation off Colonsay, West Scotland

may contain occasional arctic or boreal ostracod valves. Such mixing of fauna can be difficult to resolve, but relative frequencies, state of preservation, and the presence or absence of juvenile growth stages indicate that most of the arctic ostracod records are probably reworked from older deposits.

Offshore, in the broad area of Forth Approaches and South Forties (Fig. 1), the same units become part of the bewildering sequence of channel deposits (the Forth Beds of IGS studies, *e.g.* Thomson, 1977) fortunately underlain by what seem to be extensions of the Errol Clay of the mainland (St. Abbs Beds, Thomson, 1977). As the whole succession is of units which speak of erosion, transport and time-gaps which are difficult to assess in precise terms, in contrast to the west coast basins, it would be unwise to set up any Lateglacial standard in this area without the prospect of much fuller research and sample recovery. The prospects would seem better in the area of the Moray Firth where several sequences have been established by the Institute of Geological Sciences, inshore and offshore, down to the upper surface of the last till deposited in this area. Details of ostracod fauna published for boreholes in the Cromarty Firth (Peacock, 1974) fail to reveal 'arctic' indicator species to support the interpretation of cold-climate periods, but further work is in progress for other sequences less influenced by local inshore conditions.

Of North Sea deposits proper, the conclusions recently offered seem to confirm that the greater thickness of Pleistocene in the Forties area is probably no older than Devensian, including glacio-marine units which are probably the equivalents of onshore tills (McCave *et al.*, 1977; Eden *et al.*, 1978). From accumulated ostracod evidence, no trace has been found of any marine or freshwater Hoxnian or Cromerian deposits either in substance or on the basis of reworked isolated specimens, on which grounds a Devensian age seems to be borne out. Work on benthonic foraminifera from borehole sequences of Frigg Field (Løfaldli, 1973) has effectively demonstrated a number of arctic horizons interspersed within horizons of boreal or temperate character. There is little doubt that ostracod fauna would confirm such a stratigraphy if a parallel investigation were to be made. It is perhaps worth mentioning at this point, however, that in a case such as this, while the water temperature change could be confidently expected to emerge, it is less certain that 'older' or 'younger' arctic horizons could be distinguished. This stems from the fact that we have little present evidence for the evolution of new species of ostracod, or the extinction of older Pleistocene species within the space of the last 50,000 years. Thus no clear indicator ostracod species occur with which one could distinguish between the three arctic horizons of the St. Abbs Beds, the Swatchway Beds or the still older Aberdeen Ground Beds. Once again, the contrast between this situation and the confident precision as to time of the Atlantic core correlations must be apparent.

European Shorelands of the North Sea

In southern Scandinavia, including Denmark, whilst Lateglacial foraminiferan faunas have been extensively studied, and employed in both climato- and chronostratigraphical work through the researches of Feyling-Hanssen (1971), Knudsen (1971), and Andersen (1971), there is a sad absence of information about the ostracods from the same sections. The coast of Jutland offers splendid clay cliff sections, which embrace the Weichselian (Hirtshals), Lateglacial (Norre Lyngby) and Flandrian (Løkken) in a fashion which is in marked contrast to the exposures available on the opposite shores of the North Sea. Again, in Norway clays from the west coast south of Stavanger have been mentioned in accounts of the foraminiferan fauna (Feyling-Hanssen, 1971) and more recently, other Lateglacial sections have been described from Ågotnes (Mangerud, 1977), but in neither case do we have information relating to the ostracod fauna. The documentation of the fauna from any of these sequences would be a vital step towards any attempt to correlate between Britain and north-west Europe.

As has been pointed out in several accounts of North Sea geology, there is a marked contrast in Pleistocene successions north and south of the Dogger Bank (Caston, 1977; McCave *et al.*, 1977). To the north, the greater part of the sequence is Devensian. To the south, older sequences occur, involving Hoxnian, Cromerian, and preglacial Pleistocene successions in the areas separating East Anglia and the Netherlands. Once again, little information regarding ostracod fauna has been published for the southern North Sea (Oele, 1968, 1971), so preventing logical extensions from the Hoxnian Nar Valley Clay, or the Cromerian Forest Bed complex towards equivalent age Holsteinian and Cromerian deposits in the Lower Rhine Basin. Here too, much work remains to be done before we can attempt the exercise of correlating with that important quarter of the Continent.

CONCLUSIONS AND PROSPECTS

Anyone could be forgiven the conclusion on the evidence presented so far that ostracod fauna presents very little prospect for improving any chronostratigraphy for the Lateglacial, although their value in assessing climatic changes might be granted. This would be too gloomy a view, for certain lines of enquiry may improve the situation in due time. For example, in the past, ostracod workers have probably accepted too readily that there has been no evolutionary change at the species level over the past 50,000 years. Recently, work done upon the genus *Cytheropteron* taken at close intervals through long cores from Forties Field, has revealed variation in characters which could merit the distinction of several species where previously only one had been assumed (Masson, pers. comm.). Such discoveries are not uncommon these days when we find the scanning electron microscope revealing subtle, fine detail patterns and structures in ostracod valves of which we were quite unaware from conventional optical microscope studies. Progress may come then from improvement in our present understanding of ostracod taxonomy.

What is equally essential, is that we need to establish more precise details of the ecological distribution of ostracods on the Continental Shelf today. To this end the present author, with some members of the Ostracod Section of the British Micropalaeontological Society, is engaged in the long-term study of grab-samples provided by the Offshore Units of the Institute of Geological Sciences, with the aim of building up a simple distribution map for live species of British waters. Only as a result of this kind of work, recognising the diversity of modern fauna, its variation with water temperature and in relation to differing sediment substrates, will we be able to be confident in our palaeoecological interpretations of the Lateglacial in the areas discussed above.

Finally, any review such as this quickly reveals geographical areas for which very little data seems to be available to be introduced into discussions. In some cases, when valuable sections have been available in boreholes and studied for a wide range of organisms, ostracod studies have not been included in the total programme of work. In other cases, there is a need to return to the very full record of Scottish sites originally described by Brady, Crosskey and Robertson, and to set about a renewed examination after the lapse of a century to see what the true significance might be for the shoreland stratigraphy of Aberdeenshire, Buchan or Argyll. Ultimately, however, we inevitably come back to the need for a series of boreholes specifically planned to set up standard profiles for the Lateglacial in all its contrasting facies, preferably backed up with as many radiocarbon dates as can be determined for particular events. Otherwise, if this contribution seems to have been protesting that 'the time is not yet ripe' for ostracod correlation of the Lateglacial, at least it may have defined some of the problems involved.

Note: No attempt has been made in this review to illustrate the ostracod species discussed and interpreted in the text. Those interested in following up this aspect are referred to Brady, Crosskey and Robertson (1874), and Robinson (1978), the latter illustrated with stereoscan plates, and offering further references to the literature of ostracod taxonomy.

Problems Associated with Radiocarbon Dating the Close of the Lateglacial Period in the Rannoch Moor Area, Scotland

J. J. Lowe and M. J. C. Walker*

(City of London Polytechnic)
**(St. David's University College, Lampeter)*

ABSTRACT

Radiocarbon dates spanning the Lateglacial-Flandrian transition are presented from seven sites in the Rannoch Moor area of the Grampian Highlands, Scotland. Problems associated with the interpretation of these dates are discussed, and possible implications for the dating of the close of the Lateglacial period in other areas are then considered.

INTRODUCTION

A number of proposals have recently been put forward for a Pleistocene-Flandrian (Holocene) boundary (*e.g.* Nilsson, 1965; Morrison, 1969; Mercer, 1972; Mörner, 1976), but the recognition of a boundary stratotype of global significance is complicated by the time-transgressive nature of palaeoenvironmental changes. Within the context of NW Europe, definitions of the Pleistocene-Flandrian transition have been related mainly to terrestrial sediments of limnic origin and associated pollen-stratigraphies, with the boundary being placed where there is clear biostratigraphic evidence for climatic amelioration. A Late Weichselian-Flandrian nomenclature was formalised by Mangerud *et al.* (1974), although this has now been modified with the proposal that the term 'Holocene' should replace 'Flandrian' (Mangerud and Berglund, 1978); some authors still argue, however, for the retention of the term 'Flandrian' (*e.g.* Hyvärinen, 1978). The base of the Flandrian/Holocene is placed at 10,000 BP. A scheme proposed for northern Britain (Pennington, 1975a) differs in detail for the Lateglacial, but also places the Lateglacial-Flandrian boundary at 10,000 BP.

Gray and Lowe (1977) have discussed some of the problems associated with the stratigraphy of the Lateglacial and the Lateglacial-Flandrian transition in Scotland, particularly those difficulties which arise over radiocarbon dating. They questioned the validity of defining zone boundaries for the Lateglacial, and advocated caution in attempting correlations not only on an international scale, but even on a regional basis. They drew attention specifically to the lower boundary of the Flandrian, and noted that there was evidence to suggest that the boundary post-dated significantly the major thermal improvement at the close of the Loch Lomond Stadial (Younger Dryas). Indeed, they observed that should further early radiocarbon dates be obtained from biostratigraphic boundaries marking the end of stadial conditions, it may be necessary to follow Pennington (1977b) and subdivide the Loch Lomond Stadial into an early cold phase up to *c.* 10,400 BP and a later warmer period up to *c*. 10,000 BP. Clearly, such a step would have major implications for the location of the boundary marking the end of the Lateglacial in Scotland, and could cause serious problems for chronostratigraphic correlations between Scotland and other areas of NW Europe.

There is, however, a more fundamental problem which affects any attempt to establish regional stratotypes and chronologies (at least in areas of recently deglaciated terrain), and that is the difficulty involved in assessing the reliability of the radiocarbon dates upon which chronostratigraphies are based. This is particularly acute when there is a possibility of hard-water error resulting from the incorporation of unweathered glacial rock flour into the dated sample, and the magnitude of error stemming from this and other sources is not easy to establish. The purpose of this paper is to highlight these problems by examining a number of radiocarbon dates that have been obtained in recent years from basal limnic sediments in kettle-holes in the Rannoch Moor area of the Grampian Highlands of Scotland. The wide variation in the radiocarbon ages of a number of these samples demonstrates the difficulties inherent in dating the close of the Lateglacial period (the Lateglacial-Flandrian transition) within this very restricted geographical area, and underlines the serious problems that exist in attempts to establish broader chronostratigraphic schemes.

BACKGROUND TO THE RANNOCH MOOR PROJECT

Since 1974, the authors have been engaged in a project designed to date the disappearance of Loch Lomond Advance ice (Sissons, 1976b; this volume) from Rannoch Moor and surrounding areas. The Loch Lomond Advance is equated with the Younger Dryas Readvance of continental NW Europe. Rannoch Moor (Fig. 1) is a vast, blanket bog-covered moorland, that stretches *c*. 25 km from near Loch Tulla in the south to Corrour in the north, and *c*. 20 km from the upper reaches of Glen Coe in the west to Loch Rannoch in the east. The floor of the Moor, which lies at an altitude of *c*. 300 - 350 m, is underlain by granite, and is surrounded by mountains (locally exceeding 900 m in altitude) of complex folded metamorphic and igneous rocks, through which a system of deep valleys radiate away from the central moorland area.

This radial system of valleys reflects the fact that throughout the Pleistocene Rannoch Moor was a major centre of ice dispersal (Sissons, 1967). Thus, at the maximum of the Loch Lomond Advance, ice flowed out from the corries in the encircling mountains to coalesce on the floor of the Rannoch basin, and thence by way of the major exit valleys to the lower areas beyond. The ice-cap that developed over Rannoch Moor during this last glacial phase was *c*. 450 m thick in the vicinity of the upper reaches of Glen Coe (P. Thorp, unpublished), and major ice streams flowed along the valleys of Lochs Treig and Ossian (Sissons, 1979b); of Loch Rannoch, Glen Lyon and Glen Dochart (Thompson, 1972); of Glen Orchy (D.G. Sutherland, unpublished); and of Glen Etive, Glen Coe, Loch Leven and Glen Nevis (P. Thorp, 1979 and unpublished).

The melting of the Loch Lomond Advance ice left extensive deposits of hummocky moraine that now cover Rannoch Moor and the lower slopes of the major outlet valleys. Between the mounds, deep kettle-holes are relatively scarce, but systematic field work over several years has revealed twelve enclosed basins in which limnic sediments and telmatic or terrestrial peats have accumulated (Fig. 1). The basal deposits in all of the basins would have accumulated after Loch Lomond Advance ice had virtually disappeared from Scotland, for it seems reasonable to assume that ice would waste from the centre of dispersal at a late stage (Walker and Lowe, 1979b).

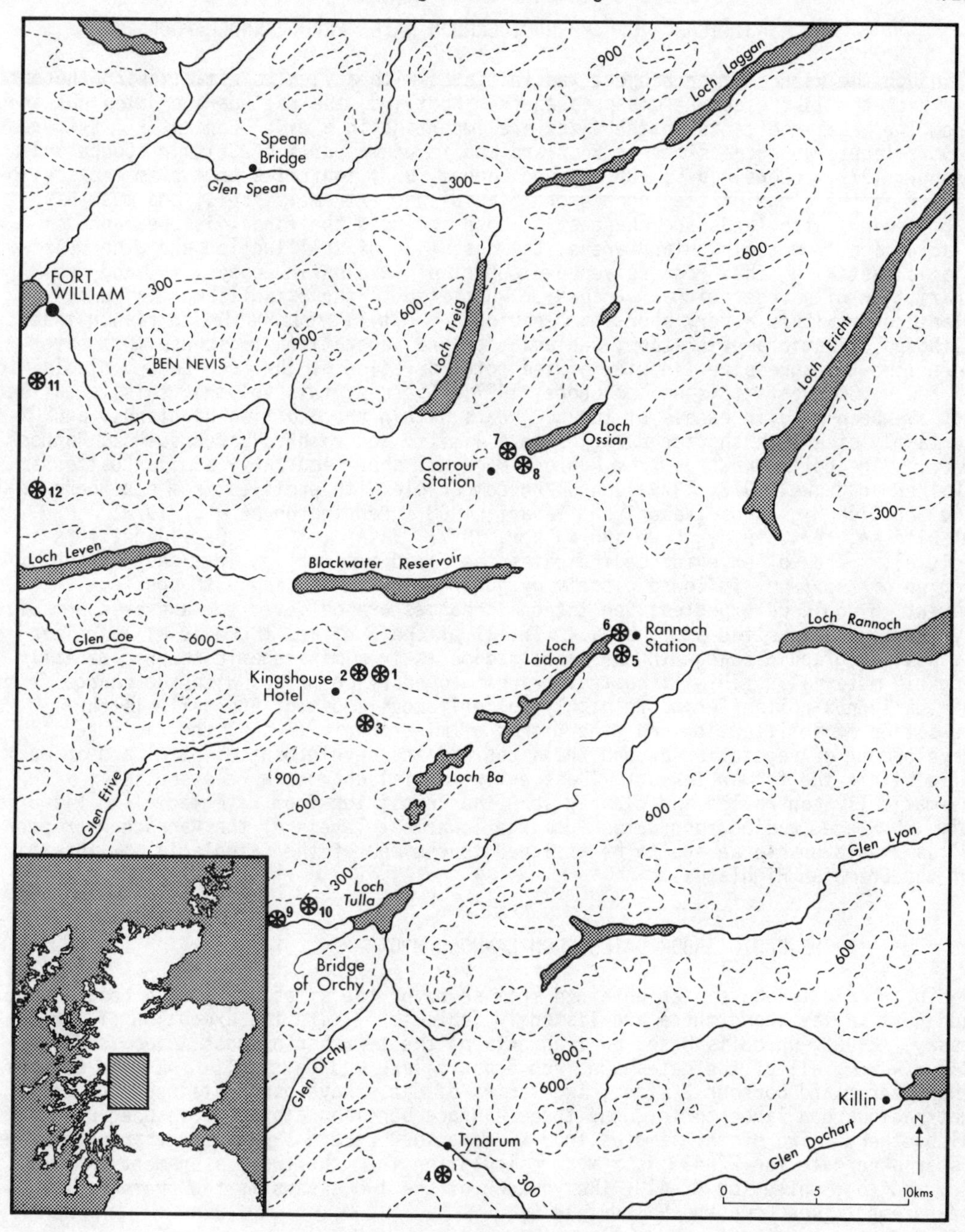

Fig. 1 Location of sites in Rannoch Moor and adjacent valleys

Contours in metres

1. Kingshouse 1	2. Kingshouse 2	3. Kingshouse 3
4. Tyndrum	5. Rannoch Station 1	6. Rannoch Station 2
7. Corrour 1	8. Corrour 2	9. Clashgour 1
10. Clashgour 2	11. Lairigmor 1	12. Lairigmor 2

THE SIGNIFICANCE OF THE RADIOCARBON DATES FROM RANNOCH MOOR

Although the Rannoch Moor project was initiated with a view to establishing the date at which the last glaciers disappeared from Scotland, the radiocarbon dates obtained from the basal sediments in the sites are important in a wider context. Evidence from coleopteran successions in Scotland and in other parts of Britain (Coope and Brophy, 1972; Coope, 1975; Bishop and Coope, 1977) indicates that climatic amelioration at the close of the Loch Lomond Stadial was extremely rapid, and must have prompted dramatic landscape changes. These included the final disappearance of glacier ice from the highland areas, the cessation of solifluction and other periglacial activity, the progressive development of lake basin ecosystems, and the initiation of a vegetational succession which led to the establishment of a closed plant cover within a very short time period. It is reasonable to assume that, although climatic amelioration, deglaciation and vegetational response must have been time-transgressive, in view of the total duration of the Loch Lomond Stadial of *c.* 800 - 1,000 years (Bishop and Coope, 1977; Sissons, this volume), this is unlikely to have been much in excess of 200-300 years within the Scottish Highlands, and certainly of a much shorter duration in a limited geographical area such as Rannoch Moor. The pollen records from Rannoch Moor all show remarkably similar patterns (Walker and Lowe, 1977, 1979a), and are comparable with profiles from elsewhere in the Scottish Highlands (Vasari and Vasari, 1968; Pennington *et al.*, 1972; O'Sullivan, 1974, 1976; Lowe and Walker, 1977; Vasari, 1977; Caseldine, this volume). The pollen evidence indicates the rapid immigration into the Rannoch Moor region of *Empetrum*, followed closely by juniper and birch, and although some local variation would be expected, vegetational changes are believed to have been broadly synchronous across the study area. In all of the profiles there is also a clear lithostratigraphic change in the basal sediments from minerogenic to predominantly organic material. This lithostratigraphic boundary coincides with the change from virtually non-polleniferous to highly polleniferous sediment and reflects the cessation of solifluction and other forms of minerogenic inwash, the rapid seral development of vegetation around the sites and the development of plant and animal life within the former lakes. These environmental changes were essentially climatically-controlled and clearly span the transition from Lateglacial to Flandrian time. Hence, radiocarbon dates from the lowermost levels of the Rannoch Moor profiles should enable an age to be assigned to the end of the Lateglacial in this part of the Grampian Highlands.

THE RADIOCARBON DATES FROM RANNOCH MOOR

Radiocarbon dates have been obtained from seven of the sites on Rannoch Moor and in adjacent valleys, and these are listed in Table 1. With the exception of the one assay, K2-433, which is based on fragments of the terrestrial moss *Rhacomitrium lanuginosum*, all of the dates are from organic lake mud (gyttja). In the case of the Tyndrum and Corrour 2 sites, two series of dates have been obtained from different radiocarbon laboratories, and three radiocarbon laboratories have been used altogether. Two of the samples from Kingshouse 2 were assayed more than once; the asterisked date for K2-433 is a mean calculation from the two measurements marked (a) and (b), while for K2-410, the younger of the two assays is the preferred measurement (see Lowe and Walker, 1976).

A minimum of two dates was obtained from the basal sediments at each site. With the exception of Kingshouse 2, where the stratigraphy is more complex (Lowe and Walker, 1976), the lowermost sample for dating was taken from the contact between virtually non-polleniferous minerogenic sediment and polleniferous organic mud. The dates relating to horizons characterised by *Empetrum* or *Empetrum-Rumex* (labelled 'pre-*Juniperus* phase' in Table 1 and Fig. 2) associations are considered to be particularly important for these assemblages are believed to reflect a very early stage in vegetational succession following climatic amelioration and ice wastage. The upper

SITE	BIOSTRATIGRAPHIC HORIZON	MATERIAL	SAMPLE THICKNESS (cm)	SITE REF.	LAB. REF.	DATE (C14 YRS. BP)	C^{13} PDB
Kingshouse 2	main *Juniperus* phase	Organic lake mud	3.0	K2-410	BIRM-724	10,370 ± 290	-21.54
						* 9,910 ± 200	-23.30
	Juniperus-Empetrum	Moss fragments	5.0	K2-433	BIRM-722	10,420 ± 230 (a)	-22.45
						10,160 ± 280 (b)	-22.45
						*10,290 ± 180	-22.45
	Juniperus-Empetrum	Gyttja	1.0	K2-436	BIRM-723	10,520 ± 330	-22.37
Corrour 2 (Series 1)	end of *Juniperus* phase	Organic lake mud	3.0	CO2-440	BIRM-854	9,800 ± 160	-24.40
	main *Juniperus* phase	Gyttja	3.0	CO2-457	BIRM-855	9,440 ± 310	-17.50
Corrour 2 (Series 2)	main *Juniperus* phase	Fine-detritus gyttja	3.0	CO2-445	SRR-1418	8,920 ± 80	-25.70
	pre-*Juniperus* rise	Gyttja	3.0	CO2-460	SRR-1419	9,440 ± 70	-29.90
Rannoch Station 1	main *Betula* phase	Telmatic peat	1.5	RS1-557	SRR-1072	9,697 ± 90	-33.40
	main *Juniperus* phase	Telmatic peat	4.0	RS1-558	SRR-1073	9,152 ± 95	-33.00
	pre-*Juniperus* phase	Gyttja	3.0	RS1-560	SRR-1074	10,660 ± 240	-31.60
Rannoch Station 2	main *Juniperus* phase	Organic lake mud	3.0	RS2-444	BIRM-859	10,160 ± 200	-24.00
	pre-*Juniperus* rise	Organic lake mud	2.0	RS2-446	BIRM-858	10,390 ± 200	-22.00
Tyndrum (Series 1)	end of *Juniperus* phase	Organic lake mud	3.0	TYN-515	BIRM-857	8,120 ± 140	-26.30
					BIRM-857	8,180 ± 110	-26.30
	main *Juniperus* phase	Silty gyttja	3.0	TYN-529	BIRM-856	8,340 ± 160	-22.90
Tyndrum (Series 2)	end of *Juniperus* phase	Organic mud	4.0	TYN-510	SRR-1416	8,040 ± 50	-27.50
	pre-*Juniperus* rise	Gyttja	3.0	TYN-528	SRR-1417	8,130 ± 40	-25.80
Clashgour 1	*Juniperus-Betula* transition	Organic lake mud	3.0	CGR-422	GU-1100	9,595 ± 215	-29.70
	main *Juniperus* phase	Silty organic mud	2.5	CGR-438	GU-1101	9,730 ± 180	-24.30
Lairigmor 2	main *Juniperus* phase	Gyttja	2.5	LM-809	GU-1083	11,300 ± 245	-29.30
	pre-*Juniperus* phase	Gyttja	2.5	LM-811	GU-1084	11,350 ± 285	-28.30

TABLE 1 DETAILS OF RADIOCARBON DATES AND SAMPLE CHARACTERISTICS

For explanation of asterisks and (a) and (b) see text. The figures in the 'Site Ref.' column give the depth in cm of the lower horizon of each sample below the present-day surface at each site. BIRM = Birmingham University; GU = Glasgow University; SRR = Scottish Universities Research and Reactor Centre, East Kilbride.

dates in the profiles were taken from levels characterised by *Juniperus* or *Betula*, and the abstraction of these samples is considered in more detail below.

The biostratigraphic horizons to which the dates relate are shown in Table 1, but these are better compared by reference to Fig. 2, where the dates for each biostratigraphic horizon are grouped together. This diagram also shows the relationship of the Rannoch Moor dates to the traditional date for the upper boundary of Jessen-Godwin pollen zone III, 10,250 BP (Godwin and Willis, 1959; West, 1977), and also the lower boundary of the Flandrian, 10,000 BP.

If it is accepted that climatic amelioration at the close of the Lateglacial Stadial and subsequent vegetational developments were not markedly time-transgressive over the Rannoch Moor region, and that the radiocarbon dating method provides meaningful age determinations within the limits of quoted standard deviations, then the radiocarbon dates listed in Table 1 should present a fairly straightforward pattern when organised according to the format employed in Fig. 2. Thus, dates on the same biostratigraphic horizon should be more or less synchronous, and there should be a consistent younging trend from the *Empetrum* phase, through the *Juniperus* phase to the period of birch dominance.

Twelve of the dates do conform to such a pattern (Fig. 2) but 8 (40% of the total) do not conform to this general trend. Some dates (RS1-558 with RS1-557, and C02-457 with C02-440) are inverted in age; there is (with the exception of the *Empetrum-Juniperus* and *Betula* dominant phases) a broad spread in radiocarbon ages for each of the biostratigraphic horizons shown in Fig. 2; and, most important of all, there is a surprisingly wide variation in the age of the lowermost samples from the seven sites ranging from 11,350 ± 285 BP at Lairigmor 2 to 8,130 ± 40 BP at Tyndrum. In view of these discrepancies in the radiocarbon age determinations, precise dating of the end of the Lateglacial in the Rannoch Moor region becomes extremely difficult, and some consideration is therefore required of the various factors that might have affected these dates and produced such contrasting results from similar biostratigraphic situations.

POSSIBLE ERRORS IN THE RADIOCARBON DATES AND THEIR INTERPRETATION

(a) Field Sampling and Laboratory Preparation

Samples for pollen analysis and radiocarbon dating were collected at the seven sites using the following methodology. The subsurface morphometry of each kettlehole was first established by traversing the bog surface systematically and taking exploratory soundings of the infill, using a level mounted at the side of the site to provide an accurate datum for individual test bores. It was assumed that the oldest material would have accumulated near the deepest point and, once that had been located, overlapping cores were removed using a piston sampler with a chamber of 5 cm diameter. The initial sequence of cores was used for pollen analysis, and additional cores were then taken from the basal levels for radiocarbon dating. The horizontal distance between any two cores was never greater than 50 cm. The cores were extruded onto clean plastic gutterpipe, wrapped in watertight adhesive sheeting, and stored in a cool room pending laboratory examination. The piston corer was completely dismantled and thoroughly cleaned between successive drives.

In the laboratory, samples were abstracted from a maximum of 5 cores and bulked to provide sufficient material for radiocarbon dating purposes. Although it is accepted that bulking of sediment may introduce problems, this procedure was considered necessary in view of the requirements to obtain thin sediment slices (see further discussion of this point in Gray and Lowe, 1977). Bulking of material is considered to be acceptable provided that there is clear lithostratigraphic or biostratigraphic

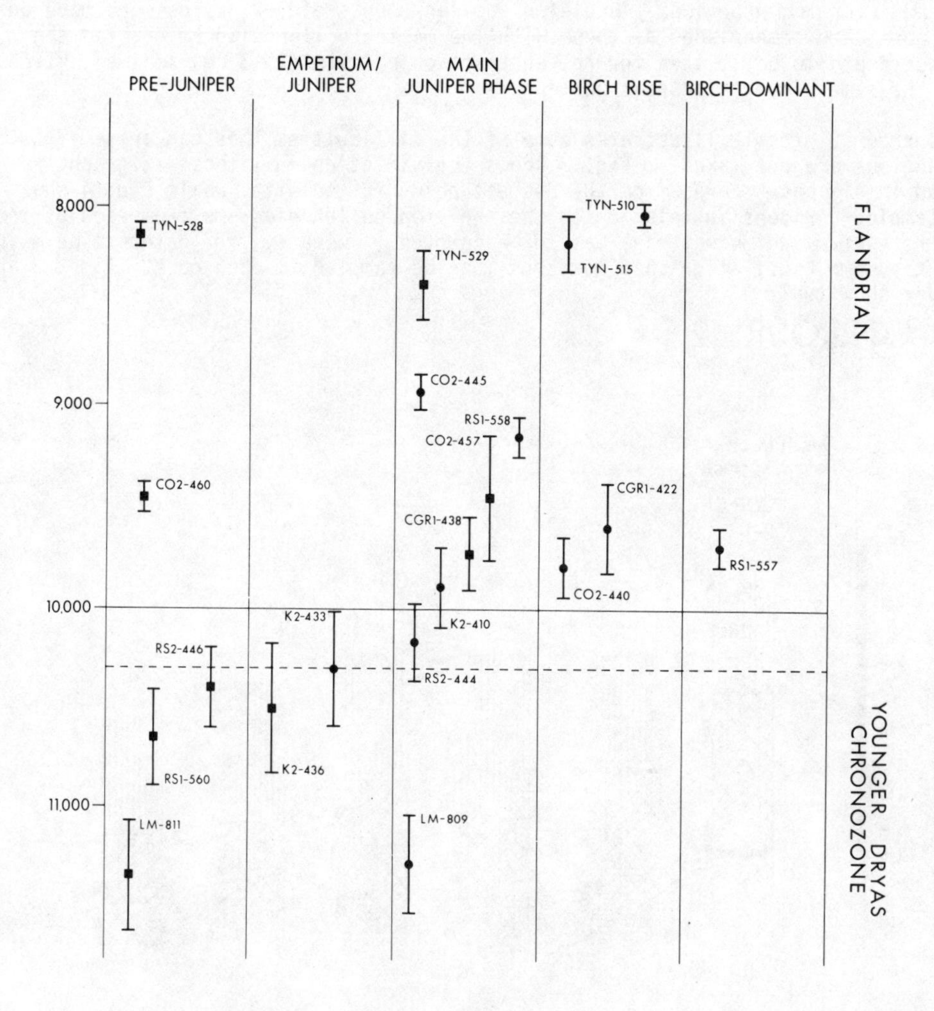

Fig. 2 Radiocarbon dates from sites in Rannoch Moor and adjacent valleys in relation to biostratigraphic horizons.

Vertical scale = radiocarbon years BP. The lower boundary of the Flandrian chronozone (10,000 BP - solid horizontal line) and the conventional age for the close of the Younger Dryas (10,250 BP - horizontal dashed line) indicated for reference. The lowermost sample obtained from each site indicated by black square.

control. The lowermost samples for radiocarbon dating from each site (with the exception of Kingshouse 2) were taken from the sharply-defined contact between minerogenic and organic sediment. At Lairigmor 2 and the Rannoch Station sites, additional dates were obtained from overlying contiguous horizons. In the Tyndrum, Corrour 2 and Clashgour 1 cores, some variation was noted in basal stratigraphy and therefore additional precautions were taken before material was removed from the cores and bulked for dating period. 'Skeleton' pollen counts of 100 grains were made on each core, and comparisons of these with the master pollen diagram enabled the correct depth to be located for the abstraction of thin (2 - 3 cm) sediment slices from the same biostratigraphic horizon.

The Corrour 2 profile illustrates some of the difficulties that can arise if such precautions are not taken. Fig. 3 shows the variation in lithostratigraphy recorded in six basal cores from the deepest point in the site, while Fig. 4 shows the sampled horizons in relation to the skeleton pollen diagrams from each of the five cores from which material was to be removed. Clearly, for dates to be meaningful, the same biostratigraphic horizons must be sampled in each core. Figs. 3 and 4 however show that:

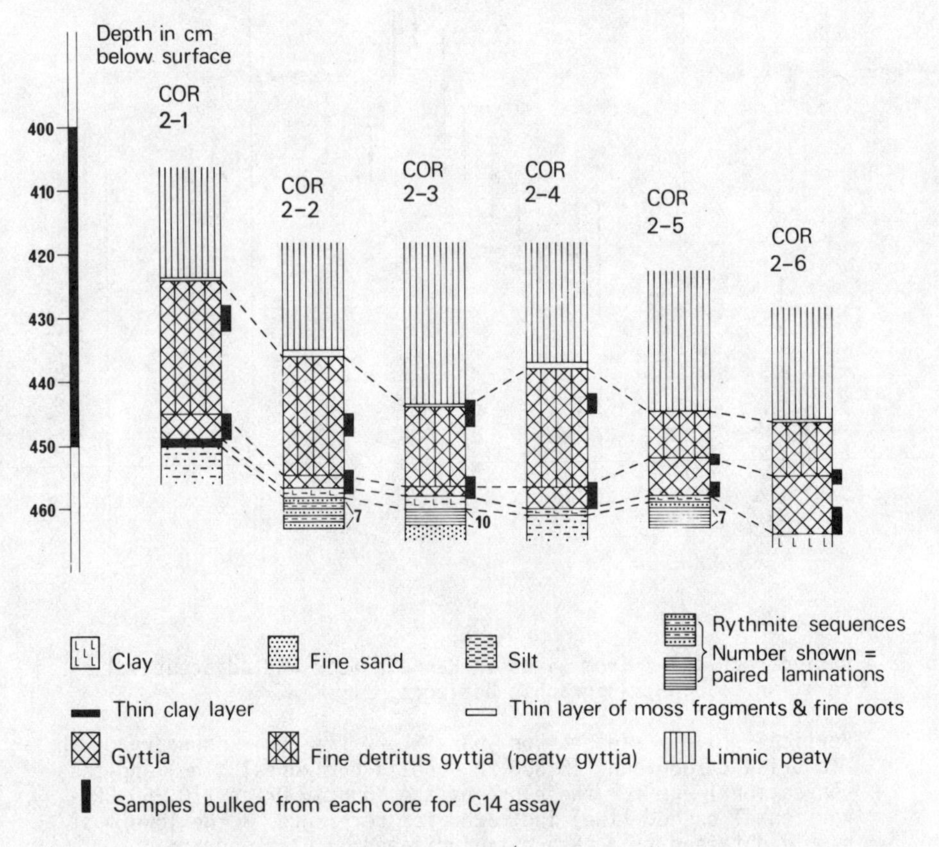

Fig. 3 Lithostratigraphy of six basal cores obtained from Corrour 2.

Dashed lines represent presumed lithostratigraphic correlations. Black shading to right = thickness of samples taken from each core and bulked for radiocarbon dating.

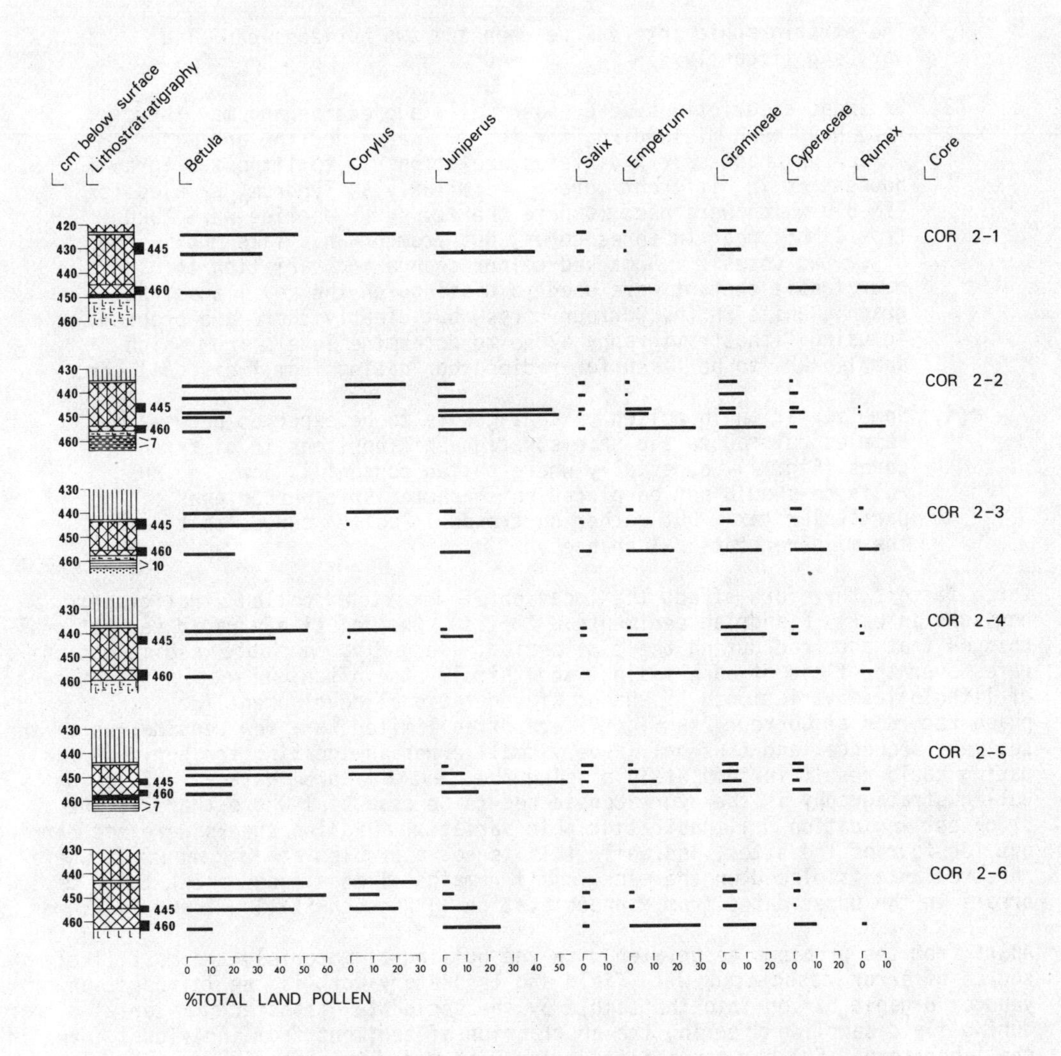

Fig. 4. 'Skeleton' pollen counts from six basal cores from Corrour 2. Lithostratigraphic symbols as in Fig. 3. 445=samples abstracted and bulked for assay CO2-445; 460 = CO2-460. Maja taxa only shown.

(1) The depth of the various horizons (both lithostratigraphic and biostratigraphic) can vary significantly.

(2) The stratigraphic interval between any two horizons can also vary significantly.

(3) Sediment descriptions were essentially subjective and may therefore have been misleading, for the *Juniperus* decline and main *Juniperus* phase bear a different relationship to lithostratigraphic boundaries in different cores. (Similarly at Tyndrum, samples for TYN-510 which were used to date the *Juniperus* decline were taken from limnic peat in three cores, but from organic lake mud in the other two cores). A marked colour change and variation in plant macrofossil content were used to distinguish the key lithostratigraphic units in the Corrour cores, but clearly there are problems in using lithostratigraphy alone to determine levels from which samples are to be taken for radiocarbon dating from individual cores.

(4) Some variation in pollen percentages is to be expected between samples related to the same stratigraphic horizons in different cores (Fig. 4), especially where pollen content is low. Thus, reliance should not be placed on characteristic percentages of particular taxa, but rather on trends in pollen curves that reflect the main vegetational changes.

Three factors therefore affect the location of important pollen-stratigraphic horizons in early Flandrian sediments. First, the relatively rapid vegetational changes that occurred during the time period; secondly, variable sedimentation rates over the floor of each basin; and thirdly, the often subjective assessment of lithological variations. Phases of vegetational development (*e.g.* the *Empetrum* phase recorded at Corrour 2 - Fig. 4) are often limited to a few centimetres of the sediment sequence, and thus only a very small error in locating the horizons for dating could result in completely missing the 'event' to be dated. Checking by pollen stratigraphy is therefore considered to be essential where there is the slightest indication of lithostratigraphic variation. Pollen checks were not carried out for four of the sites, and while this is not a problem at Kingshouse 2 where more reliance is placed on the macrofossil remains of moss (see below), there may be errors in the upper dates from Rannoch Station 1, Rannoch Station 2 and Lairigmor 2.

Apart from the problems associated with the bulking of material, the most likely source of error associated with field and laboratory work is the introduction of younger organic carbon into the sample by the accidental transfer of material either during field sampling or during the abstraction of sediment from individual cores in the laboratory. Further possibilities included under this heading are errors introduced during sample preparation, or during counting in the radiocarbon laboratory. Thus one or more of these error sources could account for the apparently aberrant dates from Tyndrum, and for the inverted dates from Rannoch Station 1 and Corrour 2.

In order to test the likelihood of field or laboratory error, samples were collected a second time from two of the sites and re-dated at a different laboratory. This gave slightly contradictory results in that the second set of dates from Corrour 2 are not inverted, but in each case the basal date is 9,440 BP, albeit with different standard deviations. On the other hand, both series of dates from Tyndrum have yielded similar age determinations. Thus, although sampling and laboratory errors could have influenced some of the dates from Rannoch Moor, particularly where unexpectedly young dates were obtained, clearly other types of errors must also be involved.

(b) Problems of Groundwater Contamination

The Tyndrum dates require special consideration. The site was first sampled in 1976 and the Series 1 dates were obtained from Birmingham University. Samples for the second series of dates were collected in 1978 and these age determinations were carried out at the Scottish Universities Research and Reactor Centre, East Kilbride. There is a remarkably close agreement between the two sets of dates, and therefore field or laboratory errors are unlikely to have affected the results. It must therefore be concluded that the quoted dates give a true record of the radiocarbon activity in the basal organic material in the site. However, the dates are of the order of 2,000 years younger than the accepted age for the immigration of juniper and birch into Scotland. There must, therefore, be some other factor affecting the basal sediments of the Tyndrum profile.

The most likely explanation is that younger organic material has been introduced into the lowermost sediments. The kettle-hole is situated in hummocky moraine and as the base of the site lies several metres above the level of the nearby river flood-plain, it would seem unlikely that fluvial activity could have affected these deposits. A more probable explanation is that groundwater seepage was the contaminating agent. One way in which this might have occurred is through intermittent fluctuation of the local groundwater table during the Flandrian. At times when the groundwater table lay below the floor of the basin, water carrying younger humic acids could have filtered down through the older basal sediments. Similar suggestions have been put forward to account for unusually 'young' dates from sites in Skye (Birks, 1973) and in the eastern Grampian Highlands (Walker, 1975a).

(c) Problems Associated with the Late Melt-out of Ice

In areas of recently-deglaciated terrain, studies have shown that residual ice may remain buried in deep kettle-holes for several hundred years after the disappearance of ice from the surrounding area (*e.g.* Porter and Carson, 1971). Although there is no direct evidence from any of the radiocarbon-dated sites that late melt-out did occur, there are indications that in other Rannoch Moor kettle-holes, sedimentation may have been delayed for some time after the general climatic amelioration (Walker and Lowe, 1979a) at the close of the Lateglacial period. Thus, although unlikely, errors arising from the late melt-out of ice cannot be entirely excluded.

(d) Errors Associated with Time-transgression in Vegetational Succession

The greatest scatter of dates on Fig. 3 is associated with the main *Juniperus* phase. Of the eight dates relating to this biostratigraphic horizon, two are inverted (CO2-457 and RS1-558), one is an aberrant date from Tyndrum (TYN-529), and one is from Lairigmor 2 (LM-809). The latter date is discussed more fully in the next section. The other four dates all overlap in range at one standard deviation, but together span the period from 10,360 BP (RS2-444 plus one standard deviation) to 8,840 BP (CO2-445 minus one standard deviation). This spread may reflect the influence of one or more of the errors considered above, as well as the statistical error associated with the measurement of radiocarbon activity. In part, however, the variation may also be a reflection of the fact that juniper could have flourished longer at some sites than at others before being shaded out by the arrival of birch. Thus, while vegetational developments are believed to have been more or less contemporaneous across the Rannoch Moor area, absolute synchroneity cannot be demonstrated, and indeed is unlikely to have occurred. While it is not possible with our present information to assess the extent to which time-transgressive vegetational changes occurred, the influence of this factor must be considered when dates from similar biostratigraphic horizons at different sites are being interpreted, even when those sites are relatively closely located.

(e) Problems of Hard-water Error

Since Deevey *et al.* (1954) first demonstrated the influence of 'hard-water' on the natural radiocarbon content of material in lakes, the influence of this source of error has been clearly shown in sediments of Lateglacial age (Donner *et al.*, 1971; Shotton, 1972). Caution is therefore required when interpreting radiocarbon dates from areas of calcareous or carbonate bedrock (Karrow and Anderson, 1975), and also from sediments that have accumulated in areas of shelly till. Thus Peglar (1979) has recently reported the very early date of 10,765 ± 310 BP from the Lateglacial-Flandrian transition at a site in Caithness where the basal sediment is "shelly, calcareous boulder clay" (p.245).

Within the present study area, there are no deposits of calcareous till, and the only locality where carbonate bedrock is encountered is in the vicinity of the Lairigmor 2 site, where an outcrop of Dalradian limestone crosses the valley some 1.5 km to the south of the basin. The dates obtained from the basal sediments at Lairigmor 2 (11,350 ± 285 BP and 11,300 ± 245 BP) are older than the conventional age for the beginning of the Loch Lomond Stadial, and it therefore seems highly probable that there is a hard-water factor associated with these age determinations.

Hard-water error in radiocarbon dates may not, however, be confined to sites located on, or near, a carbonaceous substrate, for it is argued that inert carbon can be derived from any igneous and metamorphic rock through processes of glacial erosion (Sutherland, this volume). Rock flour in newly-deglaciated terrain may provide a supply of older carbon which has been released as a result of comminution of igneous and metamorphic rock to give silt and clay fractions which are easily transported in local groundwaters and filtered into lake waters, thence to be incorporated into lake sediments and biota. As all of the basal dates from the Rannoch Moor area have been obtained from lake sediments that accumulated on fresh glacial debris following the wastage of the Loch Lomond Advance glaciers, there is a real possibility that some, if not all, of the dates have been affected by hard-water error.

On the other hand, there are some indications that, if there is a hard-water factor associated with some of the age determinations, then it is of limited extent. It has been argued elsewhere (Lowe and Walker, 1976; Walker and Lowe 1979b) that sample K2-433 is of particular importance, as that date has been obtained from terrestrial moss fragments, and that the radiocarbon age cannot have been affected by hard-water to the same extent as contemporaneous lake sediments. Yet the date (10,290 ± 180 BP) is in sequence with the basal date from the profile of 10,520 ± 330 BP and the overlying date of 9,910 ± 180 BP, both of which were obtained from limnic sediments. Moreover, if hard-water error has been a factor at Kingshouse 2 and at the other dated sites, it might reasonably be expected that only very small variations in the amounts of inert carbon incorporated into the basal sediments would have led to considerable variations in the radiocarbon dates. However, with the exception of the Lairigmor 2 and Tyndrum samples, there does seem to be a measure of accordance between the basal dates within the range of one standard deviation. There is particularly close agreement between the basal dates from Kingshouse 2 and the Rannoch Station sites (Walker and Lowe, 1979a), and these age determinations are comparable with those from other Scottish sites (many of them lying outside the limits of the Loch Lomond Advance and therefore not developed on freshly-deglaciated terrain - see Gray and Lowe, 1977) where thin slices of sediment have been taken from deposits spanning the Lateglacial-Flandrian transition. Further, if hard-water error induced by the incorporation of glacial rock flour into the dated samples has been a factor, there should be a major age discrepancy between the lowermost age determinations (*i.e.* the dates obtained from the contact between minerogenic and organic sediment) and the age determinations from the overlying organic materials. In other words, it might be expected that the most serious errors would be associated with those samples with the highest minerogenic content, and that there would be a marked decline in the influence of hard-water error (and hence a marked change in radiocarbon age) as minerogenic sediments are succeeded by predominantly organic deposits. No marked 'younging' occurs between the basal and upper dates at Clashgour 1, Rannoch Station 2 or Kingshouse 2. The Corrour 2 and Rannoch Station 1 dates cannot be considered because of the problems of age-inversion outlined above.

The evidence from Rannoch Moor is therefore equivocal for although (with the exception of Lairigmor 2) there are no reasons for believing that any of the dates have been affected by hard-water contamination, it cannot be proved that none of the age determinations have been affected in this way. In particular, the possibility cannot be excluded that small amounts of older carbon may have been incorporated into the basal sediments from unweathered glacial rock flour. The implications of this phenomenon are outlined below.

DISCUSSION

The radiocarbon dates discussed above were obtained for a project designed to establish the minimum age for the disappearance of Loch Lomond Advance ice from the Rannoch Moor area and, by implication, from the Scottish Highlands as a whole. In view of the fact that it would take some considerable time to melt the extensive ice mass that developed over Rannoch Moor and adjacent valleys during the Loch Lomond Stadial, and that the dated samples do not relate to the very earliest material that accumulated in the kettle-holes (for up to 10 cm of inorganic sediment underlie the organic deposits in each site), then the dates, if correct, imply that Rannoch Moor was ice-free before 10,000 BP and perhaps even before 10,200 BP (Lowe and Walker, 1976; Walker and Lowe, 1979b). This, in turn, suggests a major climatic improvement before that date, perhaps as early as 10,400 BP which would accord with the evidence of Pennington (1977b) in Scotland and NW England. It could be argued therefore, that the end of the Loch Lomond Stadial in the Scottish Highlands should be placed at 10,400 BP and not 10,000 BP.

However, in view of the variations in the radiocarbon dates from Rannoch Moor, and the sources of error discussed above, it is suggested that such a move would be premature. To a certain extent, errors in dating associated with field and laboratory contamination, groundwater seepage, late melt-out of residual ice, and problems associated with the time-transgressive nature of vegetational change can all be resolved by accumulating a large number of dates from adjacent sites, so that atypical and clearly aberrant dates can be isolated. A more fundamental and insidious problem affecting the Rannoch Moor dates, however, is hard-water error, for it is difficult to establish which, if any, of the dates have been so affected, and what correction is necessary to counter that error. It has been suggested by a number of authors that δC^{13} values can give an indication of those samples in which there is a hard-water factor, but as has been pointed out elsewhere (Walker and Lowe, 1979a; Sutherland, this volume), this is far from reliable when gyttja is the dating medium. The data from Lairigmor 2 underline this point. The presence of limestone bedrock in close proximity to the site, and the fact that the dates are almost 1,000 years older than those from comparable biostratigraphic horizons at other sites strongly suggests that there is a hard-water factor associated with the age determinations. Yet, the δC^{13} values (-28.3‰ and -29.3‰) are well within the 'normal' range observed for gyttja at other European localities, *i.e.* δC^{13}_{PDB} = -25 ± 5‰ (Oeschger *et al.*, 1970), and the values certainly correspond very closely to the δC^{13} counts from the other Rannoch Moor sites. Clearly, therefore, this particular index cannot be relied upon to show hard-water error in gyttja material. Nor is a measure of the carbon content of local source rocks necessarily pertinent for, except in obvious cases where limestone or a similar rock type is outcropping, when an area has been glaciated, inert carbon (albeit initially in very small amounts) may have become concentrated in morainic material, or may have been lost to the system entirely in the production of till or glacial rock flour. It could be argued that, in view of these difficulties, attempts to date the close of the Lateglacial in Scotland should perhaps be restricted to Lateglacial profiles beyond the limits of the Loch Lomond Advance. To a certain extent however, this avoids the issue, for Loch Lomond Advance ice covered large areas of the Scottish Highlands, and it would be a less than satisfactory situation if the chronology of events at the close of the Lateglacial period in the Scottish Highlands was to ignore completely dates from inside the last ice limits.

One possible way of establishing the likelihood of hard-water error in limnic sediments would be to date several centimetres of contiguous sediment slices from the lowermost horizons in a large number of profiles. If there was no dramatic change in age-depth relationship, then it could reasonably be inferred that the hard-water effect was minimal. This procedure would, however, be extremely time-consuming, and would also be expensive. It might also be complicated by some of the other factors considered above such as the influence of variations in sediment accumulation rates and the compaction of basal sediments.

An alternative and in many ways more satisfactory approach to the problem would be to ensure that macrofossil as well as gyttja material is employed as the dating medium, as illustrated by the data from St. Bees (Coope and Joachim, this volume). Unfortunately, however, the discovery of macrofossil plant material in sediments of Lateglacial and early Flandrian age in sufficient quantities for dating purposes has been extremely rare in the British Isles. The discovery of the terrestrial moss in the Kingshouse 2 profile was extremely fortuitous, and it was largely on the basis of the dates from that site that the authors have felt able to make statements concerning the timing of deglaciation (Walker and Lowe, 1979b). However, although there is some measure of agreement between the dates from that site and those from other profiles within the present study area, the variability between some of the age determinations and particularly the concern over undetectable error sources, lead us to have reservations about using the Rannoch Moor dates to assign an age to the end of the Lateglacial in this area of the Scottish Highlands.

IMPLICATIONS

The interpretation of radiocarbon dates from limnic sediments, particularly those from recently-deglaciated areas, is a problem which has far-reaching implications for the establishment of regional stratotypes, for the erection of regional chronostratigraphic schemes, and for international correlation. The problem is particularly acute for the dating of the close of the Lateglacial period, as many of the age determinations upon which currently-accepted schemes in NW Europe and North America are based have been made on limnic sediments from newly-deglaciated areas. Quite apart from the problems associated with the time-transgressive nature of climatic and environmental change, problems which have not been fully explored in this paper, but which are also of considerable importance, we suggest that there are fundamental difficulties involved in the radiocarbon dating of sediments which need to be resolved before precise dating of the Lateglacial-Flandrian transition can be achieved. If a climatic event of such magnitude as that which occurred at the close of the Lateglacial cannot be satisfactorily dated within the limited geographical area of Rannoch Moor, are we yet in a position to assign an age to the close of the Lateglacial period in different areas of NW Europe, or indeed to effect time-stratigraphic correlations between widely separated localities?

ACKNOWLEDGEMENTS

We are grateful to the Natural Environment Research Council, the City of London Polytechnic, and St. David's University College, Lampeter for financial assistance for both fieldwork and radiocarbon dating. The use of unpublished data from Mr. D.G. Sutherland and Mr. P. Thorp is gratefully acknowledged. We would also like to thank Dr. D.D. Harkness, Mr. D.G. Sutherland and Mr. R.E.G. Williams for valuable discussions on material presented in this paper.

Problems of Radiocarbon Dating Deposits from Newly Deglaciated Terrain: Examples from the Scottish Lateglacial

D. G. Sutherland

(University of Edinburgh)

ABSTRACT

Errors that are likely to influence the accuracy of radiocarbon dates from recently deglaciated terrain are considered. Younging errors which can result from such factors as late melt-out or sediment mixing are contrasted with errors that produce an ageing in the radiocarbon dates due to the hard-water effect and to the unique chemical and hence biological character of recently deglaciated terrain. Means of detecting the presence of these errors and the possibility of their removal by laboratory pre-treatment are discussed. Although each radiocarbon-dated site must be considered on its own merits, it is concluded that current practice can neither detect nor remove all the contaminants and that radiocarbon dates from recently deglaciated terrain are likely to be in error, the most probable errors resulting in an ageing of the sample and hence too early an estimate of deglaciation. Throughout, examples are taken from the Lateglacial period in the Scottish Highlands.

INTRODUCTION

In much of the Scottish Highlands the radiocarbon chronology of deglaciation during the decay of the last ice sheet and during the disappearance of the Loch Lomond Stadial glaciers is dependent upon the dating of basal organic deposits in lakes and kettle-holes. Table 1 lists such dates relevant to ice sheet deglaciation and Table 2 the dates relevant to the disappearance of the Loch Lomond Stadial glaciers. The dates in both of these tables have been expressed as a range covering one standard deviation about the mean as it is felt that this conveys the uncertainty associated with the dates more than does the standard method of expression.

Inspection of Table 1 suggests that all the dates are in agreement, indicating deglaciation over a wide area at around 12,750 BP. Given the wide distribution of the sites (Fig. 1), however, it might be expected that a diachronous pattern would emerge, older dates occurring towards the periphery of the Highlands, younger dates towards the centre. This pattern is not apparent, one of the oldest dates (Loch Etteridge) actually occurring nearest the centre of the Grampian Highlands, thus suggesting the possibility that some of the dates are in error.

The dates relevant to the glaciers of the Loch Lomond Stadial present another difficulty for it has been widely accepted in the past that the stadial (approximately equivalent to pollen zone III) ended around 10,250 to 10,000 BP (Godwin and Willis,

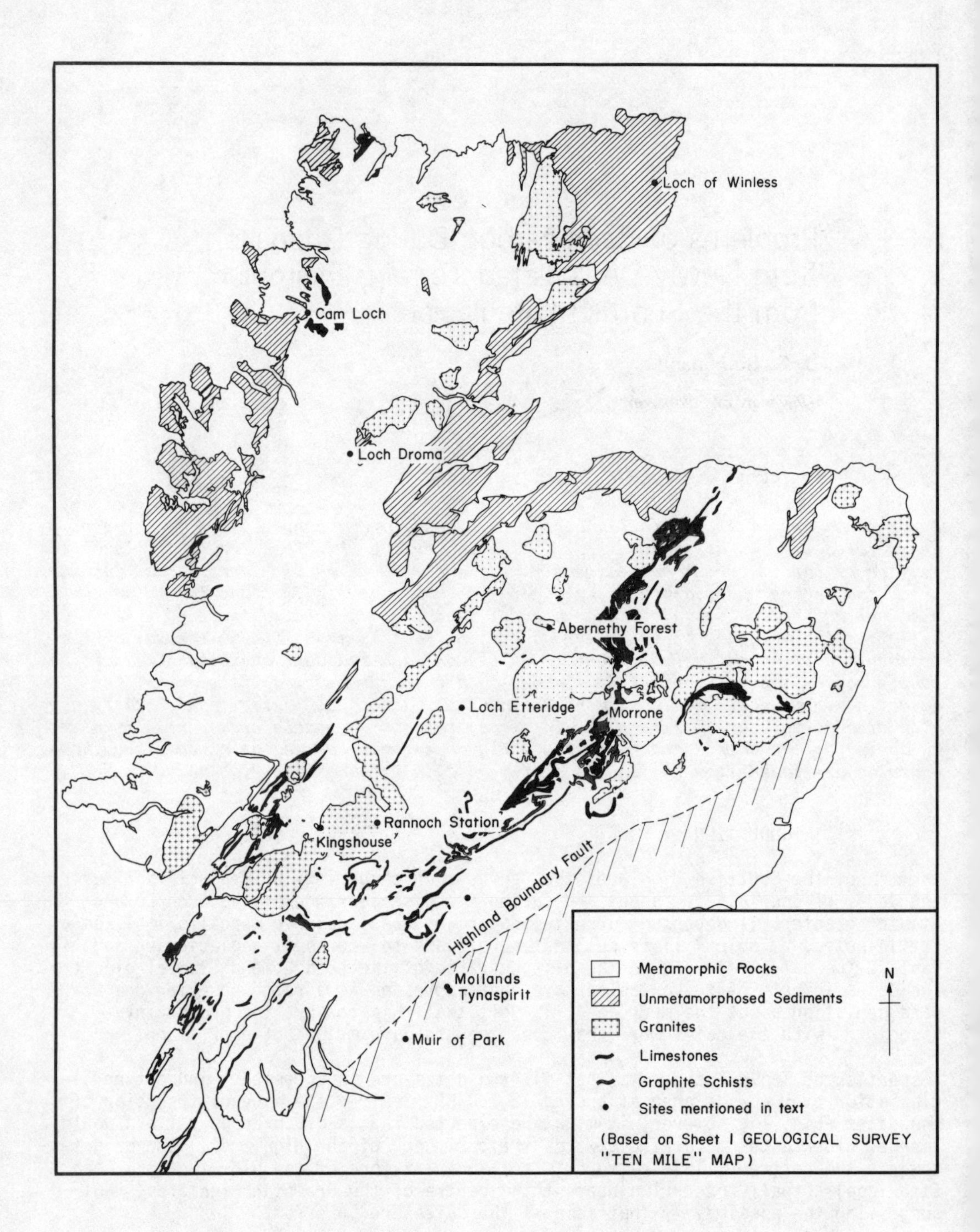

Fig. 1 Site locations and generalised geology of the Scottish Highlands

Site	Lab. No.	Age	Sample Thickness
Abernethy Forest	HEL-424	12,440 - 12,980	10 cm
Cam Loch	SRR-253	12,716 - 13,196	10 cm
Loch Droma	Q-457	12,655 - 12,965	4 cm
Loch Etteridge	SRR-304	12,761 - 13,541	3-4 cm
Loch of Winless	Q-1175	12,470 - 13,170	20(?) cm
Morrone	Q-1291	12,386 - 12,806	5 cm
Muir Park	HEL-160	12,200 - 12,820	8 cm
Tynaspirit	HV-4989	12,630 - 12,870	2 cm

TABLE 1 RADIOCARBON DATES RELEVANT TO ICE SHEET DEGLACIATION

(Sources: Vasari, 1977; Pennington, 1975a; Kirk and Godwin, 1963; Sissons and Walker, 1974; Switsur and West, 1975; Lowe and Walker, 1977)

Site	Lab. No.	Age	$^{13}C_{PDB}$
Kingshouse 2	Birm-723	10,190 - 10,850	-22.37
Mollands	HV-5647	10,585 - 10,755	-
Rannoch Station 1	SRR-1074	10,420 - 10,900	-31.6
Rannoch Station 2	Birm-858	10,140 - 10,540	-22.0

TABLE 2 RADIOCARBON DATES RELEVANT TO DEGLACIATION FOLLOWING THE LOCH LOMOND STADIAL

(Sources: Lowe and Walker, 1976, 1977; Walker and Lowe, 1979a, this volume).

1959; Pennington, 1975a; Gray and Lowe, 1977) and that the glaciers that existed during the stadial disappeared shortly after this time (Sissons, 1976b, p.106). The dates recorded for basal sediments in areas covered by these glaciers (Table 2) suggest a considerably earlier disappearance of the ice (see Lowe and Walker, this volume): hence either the established chronology or the basal radiocarbon dates are in error.

Problems associated with basal radiocarbon dates are not confined to the Scottish Highlands and certain of the errors discussed below have been considered with respect to anomalous dates in Scandinavia (*e.g.* Donner and Jungner, 1974) and North America (*e.g.* Karrow and Anderson, 1975). The purpose of this paper is therefore to examine the errors likely to affect the accuracy of basal radiocarbon dates in newly deglaciated terrain.

NATURE OF DEGLACIATED TERRAIN

Discussion of the accuracy of basal radiocarbon dates often includes statements to the effect that the dates may have been influenced by the incorporation of ancient carbon derived from the neighbouring bedrock, either by sub-aquatic photosynthesis (the well-known 'hard-water error' first documented by Deevey *et al.*, 1954), or by inclusion of carbon such as graphite in unaltered form. This section examines the possibility that newly deglaciated terrain is likely to give rise to hard-water error in basal radiocarbon dates. The effects of allochthonous carbon are discussed in a later section.

The possibility of hard-water error, it is sometimes argued, is diminished by the lack of calcareous rocks or erratics within the relevant catchment area (*cf.* Sissons and Walker, 1974; Lowe and Walker, 1976). This is an important consideration, even in the Grampian Highlands of Scotland, for limestones are widely distributed in the Dalradian rocks (Fig. 1). It must not be forgotten, however, that few rocks exist that do not contain carbon. The Handbook of Geochemistry (1966; pp6-E-1, 6-E-2, 6-M-1) gives lists of abundance of carbon in rocks. Of particular interest for the Scottish Highlands are the carbon contents of granitic and higher grade metamorphic rocks, the average contents of which are very similar, being in the case of granitic rocks an elementary carbon content of between 110 and 270 ppm C and a carbonate carbon content of between 200 and 1,100 ppm CO_2. The overall total carbon mean figure is 600 ppm (0.06%).

The likelihood of such apparently small quantities of carbon influencing radiocarbon dates depends upon (1) release of the carbon from the rocks and (2) incorporation into biogenic material. It is argued below that a glacier sliding over its bed provides a mechanism to release the carbon and hence that newly deglaciated terrain is characterised by soils and lakes in which ancient carbon is in relative abundance and available for synthesis. Evidence from both recently and formerly deglaciated terrain is cited to demonstrate that such ground is chemically and hence biologically distinct.

A temperate glacier, sliding across its bed, moves past obstacles in the bed partly by deformation of the ice and partly by pressure melting on the stoss side of the obstacles with accompanying regelation on the lee side (Paterson, 1969, p.119). This latter process results in the solution of bedrock material on the abraded stoss side where meltwater is produced followed by incorporation of the solutes into the regelation ice forming on the lee side. An increasing number of studies have been made of the chemical composition of regelation ice (Souchez *et al.*, 1973; Souchez and Lorrain, 1975, 1978; Hallet *et al.*, 1978) and such ice is found to be enriched in Na, Ca, K and Mg. An extreme example occurs in areas of calcareous bedrock, the stoss side meltwater becoming supersaturated with $CaCo_3$ and on pressure release on the lee side precipitating the excess carbonate (Ford *et al.*, 1970; Hallet, 1976) to form small flutings of calcium carbonate which are dissolved upon deglaciation over a 10 - 100 year timescale, allowing the ancient carbon to be incorporated into soils, by plants and into lake sediments.

The immediate post-glacial environment is strongly influenced by the chemical characteristics of the bare mineral soils exposed on glacier retreat. In front of glaciers presently retreating, pH values of over 8 are frequently recorded from unvegetated soils (*e.g.* Goldthwaite *et al.*, 1966; Moiroud and Gonnet, 1977), this alkalinity on bare soils not necessarily being dependent upon nearby calcareous bedrock as is specifically pointed out by Braun-Blanquet (1965). In formerly deglaciated terrain chemical analyses of basal sediments reveal relatively high values of Ca, Mg and Na at the base (Pennington *et al.*,1972). Pollen analyses of basal sediments indicate that vegetation typical today of calcareous or base-rich conditions was typical of pioneer plant communities (Gray and Lowe, 1977). The analysis of the lipid fraction of sedimentary organic matter (Cranwell, in

Pennington, 1977b) shows that basal sediments were deposited in lakes of high trophic status. Cranwell supposed that these nutrient-rich conditions resulted from leaching of glaciated rock surfaces, in addition to which, it is suggested here, a considerable contribution must initially have resulted from the glacial deposits.

Of particular interest to radiocarbon dating is Cranwell's observation that much of the (low) carbon content in Lateglacial or early Flandrian sediments is derived from aquatic plant and microbial sources. As these sources of organic carbon derive their carbon from the water in which they live, they are liable to act as agents for the assimilation of ancient carbon in the water. In post-glacial sediments it is observed that the trophic status of the lakes declines with the establishment of soil and vegetation cover, and the contribution of aquatic plants and plankton to the organic component of the sediments decreases.

Basal radiocarbon dates are typically derived from sediments containing 1 - 4% carbon (*cf.* Pennington *et al.*, 1972). Olsson (1972, Fig. 6) gives a graph relating the amount of contaminant, the age difference between the sample true age and the age of the contaminant, and the error in the measured age. From this graph it can be seen that *c.* 6% contamination with ancient carbon will produce a 500 year error in an assay and *c.* 12.5% contamination with similar material would given an error of 1,000 years. Thus for a dated sample containing only 1% organic carbon to be 500 years in error, it need contain only *c.* 0.06% carbon originally derived from bedrock. This example suggests that with samples in which there are low carbon contents, even if only the average carbon concentration of granitic rocks is made available for incorporation into the sediment being dated, a serious error could result.

MELTWATER

The ice that melts to form a kettle-hole or dead-ice hollow and which provides much or all of the initial water in the kettle-hole is the basal ice of the ice mass that covered the area. Glacier ice contains atmospheric CO_2 that was trapped in the ice during firnification and is thus approximately the same age as the ice. In polar ice approximately 1 - 5 m^3 of ice is necessary to give a sample of suitable size for radiocarbon assay (Oeschger *et al.*, 1976). In temperate ice the concentration of CO_2 is more variable than in polar ice (Stauffer and Berner, 1978) due to interaction with interstitial meltwater, though Weiss *et al.* (1972) found CO_2 to be much more stable than other gases in the same ice. Even in temperate ice, then, CO_2 concentrations are sufficiently high to be of significance to radiometric dating, and various studies have dated ice from the ablation zones of temperate glaciers (*e.g.* Coachman *et al.*, 1958; Karlen and Denton, 1976).

Little is known of the chronology of development of the last ice sheet to cover the Scottish Highlands, and even less as to its mass balance, but its life time was probably of the order of 15,000 years (Mitchell *et al.*, 1973; Sissons, 1976b; Boulton *et al.*, 1977). Allowing for mass turn-over basal ice under the centre of the ice sheet (the Scottish Highlands) could be expected to be several thousand years old. The glaciers of the Loch Lomond Stadial were rather short lived, 750 years being a likely average figure, and taking mass turn-over into consideration the ice at the base of these glaciers may have been up to 500 years in age.

Upon melting the CO_2 in the meltwater would rapidly equilibriate with the atmospheric CO_2 and would seem to present no great hazard to radiocarbon assay. Melt-out, however, as discussed below, takes place over a period of time and thus the meltwater constantly adds CO_2 (which may be up to several thousand years old in the case of the ice sheet) to developing kettles. This mechanism therefore functions rather like 'old' groundwater seeping into a lake and producing a hard-water effect.

ALLOCHTHONOUS CARBON

Carbonaceous material incorporated in glacial deposits or contained within the ice itself will be released on deglaciation and washed into lakes or kettle-holes. Graphite derived from bedrock and organic carbon originating in previous interstadials or interglacials are the substances in this category most likely to influence basal radiocarbon dates, and though their contaminating effects have been realised elsewhere (Donner and Jungner, 1974) they have not been discussed in connection with radiocarbon dates from the Scottish Highlands.

Graphite is widely distributed in the metamorphic rocks of the Scottish Highlands (Strahan *et al.*, 1917) and the outcrop of the graphite schists and slates of Dalradian age is shown on Fig. 1 as an example. Graphite presents a problem to radiocarbon dating for it is a contaminant of 'infinite' age and it cannot be removed by the standard pre-treatment of samples in radiocarbon laboratories. On samples of low carbon content relatively small amounts of graphite can produce a major dating error. For example, Ostrem (1965) indicated an error of over 3,000 years for carbonaceous material from ice in a Neoglacial ice-cored moraine, the sample on later laboratory analysis having been shown to consist of only 0.07% by weight of graphite. Ostrem was able to identify the presence of graphite by microscopic examination and produced a rough quantitative estimate of the amount of graphite present by assuming that the difference in weight between the carbon removed by burning one fraction of the sample and the carbon removed by treatment of another fraction with H_2O_2 was the weight of graphite.

Organic material derived from earlier interstadials or interglacials is a potential contaminant which may be particularly troublesome in the case of glaciers of the Loch Lomond Stadial. This can be said for the glaciers were short-lived and much of their deposits would be reworked ice sheet drift which had previously been covered by soil and vegetation. This contamination would be difficult to identify, for example by pollen analyses, for the vegetation that immediately post-dated the Loch Lomond Stadial was rather similar to that which characterised the Lateglacial Interstadial (Gray and Lowe, 1977). In the case of interglacial contaminants this need not be the case (*cf.* Donner and Jungner, 1973) and a qualitative assessment of the contamination may be possible.

Allochthonous carbon younger than the sample may also be present in the form of humic acids that have percolated downwards in the sediment column, perhaps to be precipitated in the lower sediments with higher pH values. Godwin and Willis (1959) described an artificially induced example of such contamination at Moss Lake, Liverpool but the hydrology of closed-basin sediments is poorly understood and it is not clear with what frequency such an effect as this may be encountered. A laboratory pre-treatment with NaOH may be expected to remove humic contaminants.

LATE MELT-OUT

Kettle-holes are formed by the melting of buried ice blocks once the general ice cover of an area has disappeared. A certain time-lag must occur between deglaciation and initiation of sedimentation in a kettle-hole which may be several hundred years as has been documented for areas of Alaska covered by Neoglacial ice that are still experiencing melt-out today (Post, 1976).

Late melt-out of glacier ice containing CO_2 the same age as the ice and of regelation ice containing quantities of ancient carbon from the neighbouring bedrock would still result in contamination of the basal sediments of the new kettle-hole, and it would be extremely fortuitous if the level of contamination produced an ageing effect that balanced the time that had elapsed between general deglaciation and melt-out. For the radiocarbon-dated sites in the Scottish Highlands that are relevant to ice sheet deglaciation, pollen analyses are agreed on the rudimentary nature of the vegetation that characterised much of Scotland at these times (Vasari and Vasari, 1968; Pennington *et al.*, 1972; Lowe and Walker, 1977). If one of these sites had been occupied by ice for, say, 1,000 years after general deglaciation of the surrounding area and had then recorded a radiocarbon age 1,000 years too old due, perhaps, to hard-water error, it would have received a pollen rain relating not to the early period of deglaciation but rather from the more closed vegetation that characterised the later stages of the Lateglacial Interstadial.

This point is even more important during the decay of the glaciers that existed during the Loch Lomond Stadial as closed vegetation appears to have established itself very rapidly at the beginning of the Flandrian. Because of this rapid vegetational development the potential exists for defining pollen assemblage zones relating to very narrow time periods as have been established in the Lake District by Pennington (1978). Such assemblage zones can then be used as checks on whether sedimentation was hindered by late melt-out, an argument which has tentatively been suggested for a site in Rannoch Moor by Walker and Lowe (1979a).

The pollen evidence therefore suggests that for the sites relevant to both ice sheet decay and deglaciation after the Loch Lomond Stadial, late melt-out has not produced a 'younging' error of any great magnitude. This need not be surprising given the detailed fieldwork and checking of the lithostratigraphy and basal pollen content that often goes to establishing a site as worthy of sampling for radiocarbon dating (*cf.* Walker and Lowe, 1977).

THICKNESS OF SAMPLE

The low carbon content that typifies basal sediments in lakes and kettle-holes often results in samples as long as 10 cm being taken from cores in order to enable radiocarbon assays to be performed (Table 1). A date obtained in this fashion will relate to the average[1] age of the whole sample thickness and not to the base and will thus be younger than the time of initial organic deposition (see Gray and Lowe, 1977). The degree of 'younging' due to this sampling problem depends upon the rate of sedimentation. It may be expected that at a time of soil instability and little or no vegetation cover, sedimentation would have been faster than during periods of more mature soils and more nearly continuous vegetation cover.

In order to gain some estimate of the magnitude of potential error involved in the dates obtained from a specific site that relate to ice sheet decay, it is suggested that the rate of sedimentation at the same site during the Loch Lomond Stadial is the best available approximation to the rate of sedimentation following deglaciation. Both times were characterised by open vegetation communities and unstable soils. An immediate difficulty with this approach is the length of time assigned to the (lithostratigraphically defined) Loch Lomond Stadial, for in the present context it will be site-specific, depending on altitude, nature of the surrounding slopes and other such factors. A figure of 750 years is assumed realizing that in some instances this may be too long and, in others, too short, though general order of magnitude estimates should still be valid. Table 3 below summarizes the relevant data.

---

[1]Strictly speaking not an arithmetic average since radiocarbon decay is exponential, thus weighting the age towards the younger section of the core.

SITE	RATE OF LLS SEDIMENTATION (mm/yr)	BASAL DATE (yr BP)	THICKNESS OF BASAL SAMPLE (mm)	PERIOD OF ACCUMULATION (yr)
Abernethy Forest	0.67	12,710 ± 270	100	150
Cam Loch	0.26	12,956 ± 240	100	385
Loch Etteridge	0.17	13,151 ± 390	30-40	175-235
Muir Park	0.40	12,510 ± 310	80	200
Tynaspirit	0.31	12,750 ± 390	20	65

TABLE 3 ESTIMATED PERIODS OF ACCUMULATION OF BASAL SAMPLES

The figures suggest that the dates are from 50 to 200 years (approximately half the time of accumulation) younger than the very base of the sample, such a figure being less than the standard deviation associated with the dates.

If the argument pursued above is correct in principle, then graphs of sediment depth against age should show a quickening in the rate of sedimentation towards the base. The influence of sediment compaction is likely to be critical as a minerogenic sediment with a small organic component is likely to suffer less compaction than is the more organic sediment that often characterises the periods of fuller vegetational cover during the Lateglacial. As an effect of compaction is to decrease the apparent sedimentation rate, this factor emphasises the expected contrast in sedimentation rates between more minerogenic and more organic sediments, making the more organic sediment appear to have accumulated relatively more slowly than it did. Compaction therefore increases the likelihood that basal sedimentation as indicated on an age-depth graph is faster than in the overlying sediments. It could therefore be argued that age-depth graphs that show sedimentation to be slowest in the base or that show a uniform sedimentation rate from the time of deglaciation until the Loch Lomond Stadial are evidence for error in the dates.

SEDIMENT MIXING

Bioturbation results in sediments of different ages being incorporated in the one sample. Depth of mixing may be as great as 15 cm (Lee, 1970) though is usually less than this. The importance of the effect is dependent upon the rate of sedimentation, materials of increasingly different age being intermixed the slower the rate of sediment accumulation. Where rythmites have been found, such as at Loch Etteridge (Sissons and Walker, 1974), it can be said that no disturbance has taken place and that in the more organic sediments immediately overlying the rythmites the degree of disturbance was at all times so slight as never to break the surface of the rythmites.

Sediment mixing results in older carbon being moved up the sediment column and younger carbon being moved down. In a uniform column of sediment a sample slice from the middle may be presumed to have equal amounts of young and old carbon mixed in, hence effectively cancelling each other out in the radiocarbon assay. The differences in influence on the decay rate of older and younger contaminants may be discounted for practical purposes where the contaminants are both within ±500 years of the sediment slice being assayed.

The above argument cannot be advanced for basal sediments where mixing can only result in younger material being moved in. If the basal sediments have received an initial contamination due perhaps to hard-water conditions that obtained when they accumulated, the mixing will reduce the level of this original contaminant in proportion to the amount of young carbon introduced and, conversely, the overlying layers will receive an equivalent amount of old carbon. The effect on the two layers will not, however, be equal and opposite since in the one case the contamination will be with ancient carbon while in the other it will be with carbon perhaps less than 1,000 years younger. An example best illustrates the point. Consider two layers of sediment, one overlying the other, the upper layer being 1,000 years younger than the lower (basal) one. The lower layer contains 20% ancient contamination. During deposition of the upper layer sediment mixing takes place such that 10% carbon is interchanged between the layers. This results in the upper layer having 2% ancient carbon and 8% 1,000 year-old carbon, and the lower layer receiving 10% further contamination by carbon 1,000 years younger than it. The graph in Olsson (1972, Fig. 6) allows assessment of the errors as follows. The bottom layer has an ageing error of *c.* 1,700 years due to its 18% contamination with ancient carbon and this is only slightly counteracted by *c.* 50 years younging due to its 10% contamination from the 1000 year younger layer above. The upper layer is aged *c.* 170 years by its ancient contamination and a further *c.* 50 years by its 8% contamination with 1,000 year older carbon.

This example uses figures that are large but still possible (*cf.* Shotton, 1972, who records a hard-water error of *c.* 1,700 years). The problem is one that is site-specific but in the special circumstances of newly deglaciated terrain in which there is a strong possibility of basal contamination by old carbon it seems likely that the effect of such sediment mixing as may have taken place is to age the higher layers of the sediment more than the lower layers are reduced in age. This is mainly due to the closeness in age of the overlying sediment to that under it: the faster the sedimentation the nearer in age will given thicknesses of sediment be to each other and the greater will be the potential for moving ageing contamination upwards from the base of the sediment column.

RELEVANCE OF $\delta^{13}C$ VALUES

In assessing whether a radiocarbon-dated sample has been affected by hard-water error recourse is often made to the fact that fractionation of carbon isotopes distinguishes between aquatic and terrestrially synthesised carbon. Stuiver (1975) states that $\delta^{13}C_{PDB}$ for terrestrial material ranges from -24‰ to -34‰ for non salt marsh plants in a temperate environment while the $\delta^{13}C_{PDB}$ value for aquatics such as *Nitella, Myriophyllum, Potamogeton,* and *Chara* range over values of -8.6‰ to -18.3‰. Thus Pennington (1977a) argues that $\delta^{13}C_{PDB}$ values of -10.2‰ and -17.0‰ for samples SRR-679 and SRR-680 from Lake Windermere probably are influenced by hard-water error, and conversely Lowe and Walker (1976) argue that a $\delta^{13}C_{PDB}$ value of -22.37‰ for sample BIRM-723 from Rannoch Moor suggests a lack of carbon from aquatic photosynthesis. The latter argument, however, need not be true. Firstly the $\delta^{13}C_{PDB}$ values are averages for the whole sample. Thus an organic component made up of 50% of terrestrial material at -32.0‰ and 50% of aquatic material at -16.0‰ would be assigned a $\delta^{13}C_{PDB}$ value of -24.0‰, still within the range of terrestrial material but if the lake in which the aquatics grew contained ancient carbon the sediment would be heavily contaminated for radiocarbon dating purposes. Secondly, there is a third source of organic carbon in lake sediments, that of plankton, which as was mentioned earlier, makes a considerable contribution to the organic carbon of lake sediments soon after deglaciation. Plankton has a $\delta^{13}C_{PDB}$ value close to -28‰ (Handbook of Geochemistry, 1966; Stuiver, 1975) and is thus indistinguishable on that basis from terrestrially-derived organic carbon. It therefore follows that $\delta^{13}C_{PDB}$ values may be a test for the presence in certain circumstances of aquatically synthesised organic carbon but are not evidence for its absence. A further complication is graphite which, it has been argued earlier, is a possible contaminant in basal sediments. Graphite has a large range of $\delta^{13}C_{PDB}$ values, from *c.* 0‰ to over -34‰ (Handbook of Geochemistry, 1966), which if present in a sample reduces still further the interpretive value of a measure of carbon isotope fractionation.

CONCLUSIONS

In the above consideration of the factors that may affect a basal radiocarbon date an attempt has been made to show that newly deglaciated terrain is distinctive in its chemical and hence biological character. The importance of this to radiocarbon dating is twofold: firstly that glaciation makes available 'old' carbon from a variety of sources - rocks, ancient carbon dioxide, reworked organic matter; and secondly that the lack of soil development and the sparse vegetation on newly deglaciated terrain result in lake sediments being deposited that are low in carbon content, with a high proportion of this carbon being derived from aquatic sources.

The low carbon content of the samples dated has resulted in the samples being vulnerable to relatively small total amounts of contamination. Certain reviews (*e.g.* Shotton, 1967; Bowen, 1978) have tended to diminish the influence of ancient carbon as a contaminant since the amount of ancient carbon necessary to produce a given error is considerably greater than the amount of modern contaminant that will produce the same magnitude of error (*e.g.* for a sample 12,000 years old, 10% ancient carbon produces an error of +850 years, whilst 10% modern carbon produces an error of -2,400 years). With samples, however, of low total carbon content the relatively high levels of contamination necessary for ancient carbon to affect the dates can be achieved, especially in an environment favourable for its synthesis.

If it is accepted that the character of newly deglaciated terrain is distinct from that of times of fuller vegetational development then considerable doubt must be cast on the practice of dating the upper layers of lake sediments to determine the approximate magnitude of hard-water error in the sediments (*cf.* Ogden, 1965; Karrow and Anderson, 1975). Rather it is suggested that independent tests should be made as to the likelihood of contamination of the basal samples, such as the proportion of organic carbon that is derived from plankton; the presence of graphite; or the ratio of organic carbon to total carbon.

The diverse provenance of the carbon which occurs in basal samples complicates the assessment of error in the dates. It has been pointed out that the measure of carbon isotope fractionation (necessary for the computation of technically comparable radiocarbon dates; see Stuiver and Polach, 1977) is not an unambiguous identification of aquatically synthesised organic carbon and hence this parameter can only be used in conjunction with other lines of evidence in assessing the possibility of hard-water error occurring in a dated sample. Additionally, laboratory pre-treatment of samples is complicated by frequent lack of knowledge of the origins of the carbon in the sample. Graphite, for example, cannot consistently be removed from a sample even should it be identified.

Consideration has been given to various younging factors such as late melt-out, downward percolation of younger humic acids, thickness of sample and sediment mixing. It is concluded that with suitable field and laboratory work sites affected by late melt-out can be identified; that humic acid contamination can be removed at the pre-treatment stage of sample analysis; that rates of sedimentation are likely to have been sufficiently high to diminish (though not remove) the problems associated with long samples; and that sediment mixing has probably not resulted in a major younging of the basal sediment ages. In general, it is thought that younging factors can be more adequately dealt with by existing procedures of sample recovery and analysis than can the ageing factors also discussed.

In the Introduction the unsatisfactory nature of the radiocarbon chronology of deglaciation in the Scottish Highlands was outlined. This review of the possible errors in basal radiocarbon dates has highlighted the uncertainties that surround such dates and has suggested a variety of ways in which some of the errors may be identified, if not quantitatively assessed. In view of these sources of error it is suggested that timing and extent of ice sheet decay is not accurately reflected by a series of dates isolated from each other and unrelated to relative chronologies of deglaciation. It is unfortunate that in areas in which relative chronologies do exist, such as along the Forth Valley (Sissons, 1976b, p.121) no basal radiocarbon dates are available from the numerous kettle-holes which occur in relation to various sea-levels and ice marginal positions.

Similarly the disappearance of the Loch Lomond Stadial glaciers is certain to have been a diachronous event and will only be reliably dated by carefully chosen samples that are related to each other through independent relative chronologies based on field evidence.

ACKNOWLEDGEMENTS

During the production of this paper valuable comments were made on an early draft by Dr. J.J. Lowe, Dr. J.M. Gray, Dr. J.A. Matthews, Dr. J.B. Sissons and Dr. M.J.C. Walker. In addition comments made by Dr. J. Mangerud and Dr. D.D. Harkness after a reading of the paper at the QRA Symposium in London, January 1979 were taken into consideration in producing the final version. All the errors remain the author's, however.

The Reconstruction of the Lateglacial Environment: Some Problems Associated with the Interpretation of Pollen Data

P. D. Moore

(King's College, University of London)

ABSTRACT

The potentialities of pollen analysis within Lateglacial deposits are limited by the precision of identification currently possible. Some of these limitations are discussed with respect to certain pollen taxa. Since modern counterparts to the Lateglacial vegetation are unlikely to exist at present, one must rely upon what is known of the limits of certain indicator taxa, assuming their requirements to have remained unchanged. The knowledge of current climatic limits of species is still largely based upon distributional rather than experimental evidence. Although qualitative pollen data (presence) regarding such species may be useful, the use of quantitative data as a climatic index is liable to result in oversimplification. There are considerable problems in applying specific biogeographic terms, such as 'oceanic' and 'continental', to pollen types containing a variety of taxa with differing requirements. This is especially true of the genera Artemisia and Empetrum. Only precisely defined pollen types containing species of unequivocal geographical distributions can safely be used for climatic reconstructions.

INTRODUCTION

Many problems are associated with the interpretation of pollen diagrams (see Moore and Webb, 1978) and particular caution is required in the assessment of Lateglacial pollen assemblages for a variety of reasons, which will be outlined briefly here.

The pollen analytical process consists of a series of stages, involving sampling, identification, estimation of contemporaneous species density, community composition and structure, and interpretation of the reconstructed vegetation in physical environmental terms. Sampling problems have been discussed extensively elsewhere (*e.g.* Faegri and Iversen, 1975), but particular problems of interpretation specifically related to Lateglacial pollen spectra are discussed in this paper.

IDENTIFICATION

The degree of precision with which pollen grains can be identified varies between taxa. It is advisable, however, to be aware of the precise limitations attaching to a taxon before using it as an 'indicator type' of certain environmental conditions. Very few publications of data from Lateglacial sediments include the detailed criteria upon which taxa are defined by the author and it is very probable that 'type' names for pollen taxa are being used in different ways by various workers.

Many examples of such taxonomic imprecisions, relating to taxa commonly recorded in Lateglacial pollen diagrams, could be cited:

(i) *Rumex acetosa, R. acetosella* and *Oxyria digyna*. Examination of many collections of type material of these species should be undertaken before separation is attempted. Various morphological criteria for their identification have been proposed (*e.g.* Birks, 1973) but the variation within type material is so large that a large proportion of fossil grains cannot be separated with confidence.

(ii) *Juniperus* is now counted in most Lateglacial deposits. It is not, in my opinion, separable from various other members of Cupressaceae, but these are unlikely contributors to Lateglacial pollen assemblages. More serious is the risk of confusion with certain algal and bryophyte spores which split in a very similar fashion. Isolated *Juniperus* records or low *Juniperus* percentages in pollen spectra should be viewed with caution and residual gemmae should always be sought. The problems are exacerbated by the fact that a poor state of preservation of pollen grains is commonly encountered in Lateglacial stratigraphy.

(iii) *Betula nana* is a taxon of considerable interest and importance ecologically and is very often cited on pollen diagrams without definitive characters being stated. Birks (1968) has provided useful numerical criteria for its separation and the inclusion of scatter diagrams of *Betula* measurements would be a useful addition to Lateglacial pollen papers, thus permitting the critical assessment of the data. One receives the impression that many records of '*cf. B. nana*' are based on purely subjective determinations, which are not reliable, for the criteria for separation are simply not mentioned.

(iv) *Salix herbacea*, sometimes with 'type' suffixed, is often recorded. Type material of this taxon is very variable, sometimes totally lacking the 'knotted' appearance described by Faegri and Iversen (1975). Other species, such as *S. pentandra*, may closely resemble this type and there seems little basis for its separation in the current state of knowledge.

(v) *Artemisia* is a very large genus and there is considerable pollen morphological variation both between and within species. A pollen type resembling *A. norvegica* is occasionally separated (*e.g.* Moore, 1970; Birks, 1973; Williams, 1976). A more precise definition of this type is needed, in terms of echina height and density. It is likely that some subdivision is possible within the genus on the basis of pollen morphology and some rather cursory attempts have been made (*e.g.* Webb, 1977). A great deal more work is necessary upon this genus which plays such a critical role in Lateglacial assemblages.

(vi) *Empetrum* is another genus which is regarded as valuable in Lateglacial environmental reconstruction (see below). Again, the criteria by which it is separated are rarely quoted in papers and concern must be expressed regarding the precision with which identifications are made. The separation of *Empetrum* in the key of Faegri and Iversen (1975) is not entirely satisfactory. Considerable variation occurs within type material, but the short colpae and distinctive sculpturing normally provide the basis for identification. The use of oil immersion and phase contrast microscopy is, however, absolutely essential for precise determination. Few workers separate the two sub-species, *E. nigrum* spp. *hermaphroditum* and spp. *nigrum* based on size criteria (see Andersen, 1961). The presentation of histograms of *Empetrum* grain sizes would add much to the potential of such data for interpretation.

Many other taxa, such as *Helianthemum, Hippophaë* and *Koenigia* (though confusion with *Sagittaria* is possible) are relatively easy to identify precisely. The same is true of the Saxifragaceae, but it is disturbing to see on some diagrams the term '*Saxifraga*', since the genus contains a very wide variety of sculpturing types and any identification to generic level is likely to permit further identification either to species or to a small group of species.

CONTEMPORANEOUS PLANT COMMUNITIES

It is now widely believed that there are no precise present-day counterparts to the Lateglacial plant communities. Pennington (1979) has described from Greenland an *Artemisia borealis* rich community which produces a pollen rain closely resembling that of some Lateglacial phases. It is not surprising, however, that strict equivalents are unavailable, for the existence of cold conditions at a relatively low latitude would have resulted in a climate more closely resembling an alpine rather than an arctic one. It is likely that both diurnal and seasonal fluctuations in temperature were considerably greater than those now found in arctic tundra sites.

In the light of this, Watts (this volume) makes the point that the environmental interpretation of Lateglacial times is therefore dependent upon our knowledge of the autecology of critical indicator species, rather than the general reconstruction of communities and habitats. This places an even greater demand upon taxonomic precision, as outlined above.

INDICATOR PLANTS AND ENVIRONMENTAL THRESHOLDS

Environmental reconstruction based upon indicator species rests upon two assumptions. Firstly, it assumes that the genetic constitution of Lateglacial plant populations, and therefore their physiological demands and responses, remain unchanged in their present-day equivalents. Secondly, it demands an adequate knowledge of the physical factors limiting contemporary plant distributions. The first assumption is tenuous and the second requirement is naively optimistic.

Many species at the present day contain ecotypes which vary considerably in their ecological tolerance. Modern studies in ecological genetics, in particular, have emphasized the rapidity with which populations become genetically altered as a consequence of intense selective pressures (Ford, 1976). The rapidly changing conditions of climate experienced in Lateglacial times undoubtedly led to similar selection and physiological adaptations in contemporaneous plant populations, especially among those with short generation time and high seed production (r-selected species). Even when the physical limits of distribution of a modern taxon are known, therefore, it is unwise to assume that Lateglacial populations behaved in precisely the same manner.

The short term and high amplitude climatic fluctuation of Lateglacial times naturally lead to an emphasis upon the climatic limitations of indicator plants. Information upon this subject is largely derived from circumstantial evidence in modern populations (see review by Conolly and Dahl, 1970). Often, for example, the current distributional limits of arctic-alpine species coincide with certain isotherms, usually those pertaining to summer maxima. Experimental evidence to back up these correlations is generally lacking, though this does not deny the value of such correlations as working hypotheses upon which future experimental work can be based.

The use of qualitative data (presence) as an indication of certain environmental thresholds having been crossed, is thus tentative but rational, if one assumes that the genetic status of the species is unaltered. Rather more worrying is the use of quantitative data in this way. For example, if a species is regarded as an indicator of 'oceanicity' on the basis of its temperature requirements, it cannot be assumed that a greater abundance of the species in the pollen record is an indication of greater oceanicity. This would be to neglect the large number of secondary factors which influence the abundance of a species within its potential range. Of particular importance in this respect are competitor species, drawing upon the same environmental resources, predators (see Harper, 1969), which may not merely limit a plant's population but may restrict it to those microsites which the predator finds least favourable, and also pathogens and parasites. Variations in the abundance of a species may thus be climatically determined, but indirectly so via a faunal or microbial component of the ecosystem.

The use of indicator types is at its weakest where the initial taxonomic determination is least rigorous. For example, the genus *Artemisia* has been used as an indicator of past climatic continentality. This can be justified only on the basis of a gross consideration of the requirements of the genus. Within Europe there are approximately 57 species of *Artemisia* (Tutin *et al.*, 1976) of which 2 species are probably introduced. If one considers the various geographical regions of Europe, the following figures can be contrived:

North Europe	5	spp.
West Europe	5	spp.
Central Europe	16	spp.
South Europe	28	spp. (including Alps)
East Europe	34	spp.

Obviously these divisions of Europe are imprecise and the figures thus approximate, but they demonstrate the greater richness of species in the more continental zones of Europe. It is true nonetheless that some species of *Artemisia*, such as *A. vulgaris*, have very wide distribution limits, whilst others, such as *A. norvegica* and *A. maritima*, have geographical distributions which could broadly be termed oceanic. Thus until a greater degree of taxonomic precision is possible within the genus on the basis of pollen morphology, it is a little presumptuous to apply the 'continental' label.

Conversely, *Empetrum* is a genus which is often regarded as oceanic in its climatic requirements. The European flora now contains two taxa which are now regarded as subspecies of a single species, *E. nigrum* (Tutin *et al.*, 1972). Of the two subspecies, *E. nigrum* spp. *hermaphroditum* is the more northern and montane in its distribution, being found in northern Fennoscandia and in the Alps, but it also extends to Iceland, Western Greenland and through arctic Siberia. Even *E. nigrum* spp. *nigrum* extends to western Siberia. Again, the uncritical application of a blanket term, such as 'oceanic', is ill-advised.

If indicator species are to be used effectively, precise taxonomic definitions and unequivocal geographical distributions are thus necessary. An example of such is the discovery of *Gypsophila* pollen belonging to either *G. fastigiata* or *G. repens* from Lateglacial sediments in SE Scotland by Webb (1977). Although the pollen identification cannot yet be taken to specific rank, both of the possible species have continental distributions in Europe. It is evidence of this sort which may eventually permit a more detailed reconstruction of the Lateglacial environment in both spatial and temporal dimensions.

CONCLUSION

Although it is tempting to interpret pollen data in climatic terms, the present state of knowledge, both in pollen morphology and in the biogeography and eco-physiology of modern plant populations, is currently very limited. In the light of these gaps in our knowledge a precise analysis of the changing climate of Late-glacial times on the basis of pollen data is still beyond our grasp.

The Stratigraphic Subdivision of the Lateglacial of NW Europe: A Discussion

J. J. Lowe and J. M. Gray*

(City of London Polytechnic)
**(Queen Mary College, University of London)*

ABSTRACT

Problems associated with the Lateglacial stratigraphy of NW Europe are discussed. These include the recognition of an Older Dryas climatic oscillation, the usefulness for the Lateglacial of the stratotype approach and some other standard stratigraphic procedures, the error sources and resolution of radiocarbon dating, the basis of Lateglacial chronostratigraphic schemes, and the validity of defining boundaries on continuous curves of environmental change. A general climatostratigraphic scheme for the Lateglacial of NW Europe is suggested which encompasses flexibility, simplicity, and accuracy in reflecting the uncertainty of present knowledge. The main purpose of the paper is to stimulate discussion for we agree with Mangerud and Berglund (1978) that NW Europe is a natural geographical region for a common Quaternary stratigraphic nomenclature, and there does seem to be considerable agreement on the Lateglacial climate of this part of the world.

THE LATEGLACIAL ENVIRONMENT OF NORTH-WEST EUROPE

Before commencing discussion of stratigraphic schemes for the Lateglacial[1] of NW Europe, it is relevant to outline some of the general characteristics of the Lateglacial environment and some aspects that are more contentious. There is general agreement that within NW Europe the Late Weichselian/Late Devensian glaciation was followed by a marked climatic improvement at about or sometime before 13,000 BP (Mangerud *et al.*, 1974; Coope, 1977; Gray and Lowe, 1977; Berglund, 1979) and conditions were warm enough to permit the establishment and development of soils and biota for perhaps as much as 2,000 years after that. A marked deterioration of climate followed, traditionally dated to between 11,000 - 10,800 and 10,250 - 10,000 BP (Godwin and Willis, 1959; Mangerud *et al.*, 1974; West, 1977) and termed the Younger Dryas (continental NW Europe), Loch Lomond Stadial (Britain) or Nahanagan Stadial (Ireland). Climatic conditions at this time were severe as attested by, for example, a major ice readvance or recrudescence of glaciers (*e.g.* Sissons, 1979c; Mangerud, this volume), widespread and varied periglacial activity including the

---

[1]It is realised that Mangerud *et al.* (1974) recommended the term "Late Weichselian" for the approximate equivalent of the Lateglacial in Norden. However, the term "Late Weichselian" now has a proposed wider definition (Mangerud and Berglund, 1978), and thus we have used the term Lateglacial for the whole of NW Europe.

development of ground ice, and more intense fluvial regimes (*e.g.* Sissons, 1979c; Rose *et al.*, 1979), rapid coastal rock erosion (*e.g.* Sissons, 1974c, 1978; Gray, 1978; Dawson, this volume), major changes in biostratigraphic records (*e.g.* Mangerud *et al.*, 1974; Pennington, 1977a; Gray and Lowe, 1977; Berglund, 1979; Coope and Joachim; Robinson; Lord, all this volume), and low inferred stadial temperatures based on reconstructed firn-line evidence (*e.g.* Sissons, this volume). The stadial was in turn brought to an end by a rapid climatic amelioration that heralded the start of the present interglacial, the Flandrian.

There is also general agreement that these major climatic changes were related to movements of the oceanic polar front. Today this is situated well to the NW of the British Isles in the vicinity of eastern Greenland, but at the maximum of the last glaciation it lay at the latitude of N Spain. Subsequently, however, its eastern end began to retreat northwestwards, clearing the British Isles around 13,500 BP (Ruddiman and McIntyre, 1973). However, towards the end of the Lateglacial polar waters returned to as far south as SW Ireland (Ruddiman *et al.*, 1977) bringing the return of glacial conditions, before retreating northwestwards again at the end of the Lateglacial.

These movements of the oceanic polar front can explain several aspects of the Lateglacial: the fact that the western end of the oceanic polar front remained relatively stationary through the Lateglacial while in the eastern Atlantic it was swinging back and forth (indeed the overall movement is often compared with a gate hinged at one end) can explain why the marked climatic changes of the NW European Lateglacial were not felt in N. America (Mercer, 1969; Watts, this volume); the steep temperature gradient across the front can explain the rapidity of temperature changes associated with its movement (*e.g.* Coope and Brophy, 1972; Coope, 1977); the time-transgressive nature of the movement leads to the expectation of time-transgressive climatic changes over NW Europe; and finally the fact that this movement was an oceanic one can explain why the effects were most marked in oceanic NW Europe and dwindled in impact towards eastern and southern Europe (see Watts, this volume).

Certain aspects of the Lateglacial environment are more contentious, particularly the recognition and status of a proposed short climatic deterioration (Older Dryas) prior to the Younger Dryas. The concept dates back to Iversen's (1954) work at Böllingsö in Denmark, and from time to time strong evidence is published for a pollen-stratigraphic subdivision of that part of the Lateglacial earlier than the Younger Dryas (*e.g.* Pennington, 1975a, 1977b; Walker, 1977; Caseldine, this volume, for sites in the British Isles), and it has been suggested that this subdivision may equate with or approximate to the Bölling - Older Dryas - Alleröd climatic oscillation proposed for continental NW Europe. For most sites in the British Isles, however, no such subdivision has been possible and Watts (this volume) has argued that evidence for a distinction between separate Bölling and Alleröd phases is absent from most sites in Europe and is weak or unsatisfactorily demonstrated at many sites where the distinction has been made, including Böllingsö itself. It may be suggested therefore that either climatic changes were slight, so that their effects can only be discerned at sites that occupied critical locations during the Lateglacial, or the majority of analyses have not been of sufficient detail (*cf.* Caseldine, this volume), or there is some explanation for the pollen-stratigraphic and lithostratigraphic changes other than a climatic one. It is not possible to decide this issue on the basis of information currently available.

Further, it has been known for some years now that evidence from coleopteran assemblages suggests different climatic developments during the Lateglacial to those based on vegetational records. Coope and Joachim's new data (this volume) from St. Bees, NW England, further emphasises this point. These differences are so consistently recorded at sites where both coleopteran and palynological information is available that there must be a fundamental explanation. The beetle evidence does not accord with the traditional concepts of Bölling, Older Dryas and Alleröd climatic episodes. The conflict with palynological data is probably explained by the fact that beetle populations are far more mobile, and vegetational developments lag behind major climatic shifts (*e.g.* Coope, 1975, 1977).

However, the study of coleopteran assemblages must also be viewed with some caution. Interpretations rest heavily on the assumption that present (and past) beetle distributions are limited by thermal thresholds. While there are strong arguments for accepting this premise, beetles have not been studied as thoroughly as plant species in recent years, and some modern distributional studies or experimental work are desirable. Secondly, the size of bulk samples taken for coleopteran studies (see Coope and Joachim, this volume) may have the result of masking important but small-scale biostratigraphic changes, which could, for instance, prevent the 'resolution' of an 'Older Dryas' phase (see Caseldine, this volume).

Climatic reconstructions for the Lateglacial are therefore problematic: there are differences of opinion over the trend of temperature changes throughout the Lateglacial, mainly as a result of different interpretations of biostratigraphic parameters. To add to these difficulties, it should be borne in mind that climate has different components, the two most important being temperature and precipitation. Some of the climatic inferences published in this volume by Coope and Joachim, Macpherson and Sissons, though perhaps very tentative in places, nevertheless serve to remind us of the importance of taking both components into account when explaining inferred palaeoenvironmental changes. Further, climate is only one of a number of ecological factors that affect *e.g.* beetle and plant distributions, and the ecological requirements of even a single species can be very complex (see Moore, this volume).

STRATIGRAPHIC SUBDIVISION OF THE LATEGLACIAL

Methods of Subdivision

Although not the original name, the International Subcommission on Stratigraphic Classification (ISSC) was created by the 19th International Geological Congress (Algiers) in 1952 as a subcommission of the International Commission on Stratigraphy. The culmination of over twenty years of work came in 1976 with the publication of the *International Stratigraphic Guide* (Hedberg, 1976). Preliminary versions of chapters in the *Guide* had previously been published as a series of ISSC Reports (*e.g.* Hedberg, 1970). According to the *Guide* the best known and most widely used stratigraphic categories are:

1. Lithostratigraphy - that element of stratigraphy which is concerned with the organisation of strata into units based on their lithologic character.
2. Biostratigraphy - that element of stratigraphy which is concerned with the organisation of strata into units based on their fossil content.
3. Chronostratigraphy - that element of stratigraphy which is concerned with the organisation of strata into units based on age relations.

It is recognised in the *Guide* however, that many other types of unit are widely used (Hedberg, 1976, pp.8-9).

Mangerud *et al.*, (1974) review the use of different types of units in the Quaternary stratigraphy of Norden, and to the three types described above they add morphostratigraphy (units based on morphology) and climatostratigraphy (units based on geological indications of climatic changes). The main conclusion of their paper was a proposal for a common chronostratigraphical classification of the Quaternary in Norden. A chronostratigraphic (also known as time-stratigraphic) unit is defined as being "a body of rock strata that is unified by being the rocks formed during a specific interval of geologic time" (Hedberg, 1976, p.67). It is important to realise, however, that this definition does not necessarily mean that the absolute dates of the "interval of geologic time" are known. "Thus we may speak of the chronozone of the ammonites, which would include all strata formed in the long interval during which ammonites existed, regardless of whether the strata contained ammonites" (Hedberg, 1976, p.67), and regardless of whether there are absolute dates on the time span. As the *Guide* points out absolute dates can be applied later, and are particularly useful in extending chronostratigraphic units (chronocorrelation)(Hedberg, 1976, pp.89-90).

Of particular relevance to the present paper are the proposals of Mangerud *et al.*, (1974) for the Lateglacial which they subdivide into four chronozones with boundaries and names as shown in Table 1.

Radiocarbon years BP	Chronozones
10,000	
	YOUNGER DRYAS
11,000	
	ALLERÖD
11,800	
	OLDER DRYAS
12,000	
	BÖLLING
13,000	

Table 1 Chronostratigraphic subdivision of the Lateglacial in Norden (after Mangerud *et al.*, 1974)

Pennington (1975a, p.167) claimed that pollen sequences from Blelham Bog in NW England and Cam Loch in NW Scotland "can be correlated by C14 dating" with pollen sequences in southern Scandinavia "and the same chronostratigraphic subdivisions applied. This means that though differences in lithostratigraphy and biostratigraphy indicate regional differentiation of climate and vegetation between these two regions of north-west Europe at the close of the last major glacial episode, the evidence for major environmental (*i.e.* climatic) changes is found at the same dates in both regions; these dates are those used for chronozone boundaries by Mangerud *et al.*, (1974)".

A stratigraphic concept that has also been advocated recently is the use of stratotypes, type-sites, or reference sections for defining Quaternary subdivisions (*e.g.* Mangerud *et al.*, 1974; Bowen, 1978), following recommended stratigraphic procedure (*e.g.* Hedberg, 1976, Chpt. 4). Several type-sites have been defined for the Quaternary of NW Europe (*e.g.* Mitchell *et al.*, 1973; Mörner, 1976), and for the British Lateglacial the profile at Low Wray Bay, Lake Windermere has been proposed as the "standard reference section" for the interstadial in Britain, which was therefore named "Windermere Interstadial" (Coope and Pennington, 1977; Coope, 1977; Pennington, 1977a). Coope and Pennington noted, however, that "the lower boundary of interstadial sediments may be drawn at different horizons, depending on the criteria that are adopted, whether they be lithostratigraphic, palaeoentomological, or palaeobotanical ... If the notion of interstadial is based on a climatic interpretation, two alternative views ... are given ..." (1977, p.338), according to whether the duration of the interstadial climate is defined on the basis of the beetle (Coope) or pollen (Pennington) assemblages.

As Bowen (1978, p.104) points out the term "climatostratigraphy" has several alternatives, such as "climatic units", "climate-stratigraphic units", "units based on climate change", "glacial-stratigraphic units", and "geologic-climate units". The American Code (1961) defined the last as "an inferred widespread climatic episode defined from a subdivision of Quaternary rocks" and boundaries were meant to be those of the stratigraphic unit on which it was based. Mangerud *et al.* (1974, p.113) define a climatostratigraphical unit as "a stratigraphical unit with the boundaries defined by geological indications of climatic changes". There has been much debate over the use of climate as a valid basis for stratigraphic subdivision, one of the problems being that it relies on inferences from lithostratigraphy or biostratigraphy - "we do not observe the climatic changes themselves, only the impact of the changes on vegetation, fauna, glaciers, oceans, sediments etc." (Mangerud *et al.*, 1974, p.113). Also the boundaries are likely to be time-transgressive and this is potentially more of a problem for the Lateglacial, with its greater resolution, than for earlier geological time.

In a previous paper (Gray and Lowe, 1977) the present authors argued for and constructed a climatostratigraphic subdivision of the Scottish Lateglacial. Two climatic units were recognised, *viz.* a "Lateglacial Interstadial" followed by a "Loch Lomond Stadial" and these were given the definitions repeated in the introduction to this book and discussed later in this paper.

Basis of Stratigraphic Procedures for the Lateglacial

Bowen (1978), Mangerud *et al.*, (1974) and others have recommended that standard units should be defined for Quaternary subdivisions, following recommended stratigraphic procedures (*e.g.* Hedberg, 1976). In turn this has led to much greater awareness of the need for rigour of meaning in dealing with Quaternary stratigraphy. As Bowen (1978, pp.77-80) discusses, however, there are particular aspects of the Quaternary that make it distinct from earlier geological time for which the general stratigraphic procedures have been mainly designed. In particular Bowen (p.78) lists the following special features of the Quaternary - "... its short time-span, inadequacy of its palaeontology for conventional utilization in subdivision, fragmentary nature of its depositional record, unusual stratigraphic relations caused by the influence of geomorphic situation, and the way in which it has been subdivided using climate as a standard." Although he stresses (p.80) that these special problems "are not so great as [to] invalidate in any way the stratigraphic procedures applied to pre-Quaternary rocks", he does agree that new kinds of stratigraphic units have been introduced for the Quaternary, including morphostratigraphy, soil stratigraphy, and geologic-climate units.

Vita-Finzi (1973, p.12) agrees that "The Quaternary presents many obstacles to the application of standard stratigraphical procedures", and Richmond (1959, p.663) points out that "the stratigraphic treatment of the Quaternary has differed in several respects from that accorded older rocks." Hedberg (1976, p.4) stresses that the aim of the ISSC *Guide* is only as a recommended approach - "not as a 'code'. There is no intention that any individual, organization, or nation should feel constrained to follow it ... procedures should not be legislated". The *Guide* is meant to "continually evolve in keeping with the growth of geologic knowledge" and it is recognised "that there are stratigraphic situations where hard and fast rules cannot be applied, and that often one must simply use common sense in deciding what in the long run will most effectively promote clarity, understanding and progress." At the same time, however, there is a warning (p.5) against the use of "unreal concepts, vague and cloudy terms and imprecise definitions". The Quaternary is described (p.1) as one of many "special or newly developing fields of stratigraphy" that will be the subject of future studies by the Subcommission which will issue supplements or revisions from time to time.

If the Quaternary as a whole has special characteristics that distinguish it from earlier geological time, then this is even more true for the Lateglacial. Some of its particular characteristics were outlined in previous sections of this paper, including the rapidity of the changes, their time-transgressive nature and their regional expression all of which have far higher levels of resolution than for the earlier Quaternary. The deposits are more abundant and better preserved and, unlike most of the Quaternary, they lie within the range of radiocarbon dating. There is undoubtedly a strong argument for treating the Lateglacial (and Flandrian) as a special case since much of standard stratigraphic procedure has little relevance to it. Bowen (1978) has said relatively little about the stratigraphy of the Lateglacial and Flandrian, but Mangerud *et al.* (1974, p.112) found that for their scheme some modifications of the rules for the pre-Quaternary stratigraphy had to be made. However, rather than squeezing our time period into ill-fitting clothes, it can be argued that there is a case for a fresh look at the characteristics of the Lateglacial and for what Bowen calls a "pragmatic approach" in deciding how best to treat Lateglacial stratigraphy.

Bowen (1978, p.193) has pointed to instances of rapid, "even catastrophic" changes during the Quaternary. Worsely (1977, p.217) believes that the "key to the past may not always be the present". With particular reference to the Lateglacial, Gray (1978, p.161) has stated that "It is probable ... that there are no exact modern analogues of the environmental conditions on western Scottish coasts at that time ... high-latitude coasts today have far less marked diurnal temperature fluctuations than western Scotland during the Late-glacial ...", while in a similar vein Moore (this volume, p.153) contends that "It is now widely believed that there are no precise present-day counterparts to the Lateglacial plant communities ... for the existence of cold conditions at a relatively low latitude would have resulted in a climate more closely resembling an alpine rather than an arctic one". If we are prepared to accept these comments on the basic geological premise of uniformitarianism, then the questioning, where considered necessary in the light of a careful evaluation and discussion of all the arguments, of the relevance for the Lateglacial of some stratigraphic procedures evolved for earlier geological time should cause us little difficulty.

A SUGGESTED GENERAL CLIMATOSTRATIGRAPHY FOR NW EUROPE

The suggestion that the Scottish Lateglacial can be subdivided into two climatostratigraphic units (Gray and Lowe, 1977) has received support in recent years from various parts of Europe. To a considerable extent our scheme was based on more informal ideas earlier expressed for Britain (*e.g.* Sissons, 1974b, 1976b; Coope, 1975).

Berglund (1979, p.110), although recognising a slight deterioration corresponding to the Older Dryas, asserts that in southern Sweden the traditional climatostratigraphic subdivision into two interstadials, Bölling and Alleröd, is "questionable", thus supporting the discussion by Watts (this volume). Berglund concludes that "the time interval 13,000 to 11,000 may be regarded as one more or less uniform Late Weichselian interstadial" and his Figure 7 shows a climatostratigraphy consisting of two units, an "Interstadial" and a "Stadial". The similarity between Figure 7 of Berglund (1979) and Figure 1 of Gray and Lowe (1977) is remarkable, and it must be stressed that these diagrams were produced completely independently. Although Berglund's paper referred to above was published two years after ours, an earlier version of his paper with basically the same diagram was published by Berglund in 1976 (Berglund, 1976, Fig. 16). Thus strong support is given to the view that a climatostratigraphic subdivision of the Lateglacial of NW Europe should involve an interstadial followed by a stadial. Berglund (1976, pp.52-57) finds wide support for this concept in many parts of Europe.

Our belief in the need for climatostratigraphic subdivision of the Lateglacial remains undiminished, and in this paper we suggest a general scheme (Table 2) that we hope can be applied throughout NW Europe. The remainder of this paper will discuss some problems with other stratigraphic classifications and procedures currently applied to the Lateglacial of NW Europe, before proceeding to discuss the terminology and basis of our scheme and why we have constructed it in the manner shown in Table 2.

Radiocarbon years BP	Climatostratigraphic Units
	FLANDRIAN INTERGLACIAL
10,000	
	transition
10,500	
	YOUNGER DRYAS STADIAL
11,000	
	transition
12,000	
	LATEGLACIAL INTERSTADIAL
13,000	
	transition
14,000	
	LATE WEICHSELIAN/ LATE DEVENSIAN/ LATE MIDLANDIAN MAIN GLACIAL

Table 2 Suggested general climatostratigraphic subdivision and terminology for the Lateglacial of NW Europe

PROBLEMS WITH CHRONOSTRATIGRAPHY

The ISSC *Guide* recommends that chronostratigraphic classification should be based on stratotypes. Mangerud *et al.* (1974, p.114) suggest that this practice "is applied for units of stage rank or higher" and "stratotypes may also be preferred for boundaries of units of lower rank". For the Lateglacial and Holocene, however, they proposed that definition of chronozone boundaries should be made directly in conventional radiocarbon years. They point out that the ISSC Preliminary Report on Stratotypes (Hedberg, 1970) makes it clear that definitions in terms of radiometric ages would accord with the concept of chronostratigraphical units. In the Preliminary Report the ISSC had two major objections to such definitions but Mangerud *et al.* (1974, p.114) argued that "none of these objections have any real significance for our proposal".

However, the revised version of the Preliminary Report on Stratotypes, published as Chapter 4 in the ISSC *Guide* (Hedberg, 1976, p.27), emphasises that "The stratotype ... constitutes the standard of reference on which the concept of the unit is uniquely based" and "provides by far the most stable and unequivocal standard of definition." Thus the *Guide* does not appear to lend support to the type of chronostratigraphic units proposed by Mangerud *et al.* (1974) for these are not based on stratotypes.

Similarly the American Code (1961) recommends that the upper and lower limits of all time-stratigraphic (*i.e.* chronostratigraphic) units should be defined in the rock succession at a type section. However, this is not a problem if it is agreed that the Lateglacial requires special stratigraphic procedures, and the type of chronostratigraphic units proposed by Mangerud *et al.* can be assessed in the context of how best to treat Lateglacial stratigraphy (see below). It is, however, also pertinent to point out that Richmond (1959, p.671) has argued that "Care should be taken ... not to define a time-stratigraphic unit in terms of isotopically measured time", and conversely not to apply time-stratigraphic (*i.e.* chronostratigraphic) terms to "units defined on the basis of such physical measurements".

It is important to examine how the Lateglacial chronozone boundaries of Mangerud *et al.* have been decided. They state (p.114) that they "propose only precise definitions of already well established terms" and later (p.117) that "In Denmark, southern Sweden and southern Norway ... the old climato-biostratigraphic zones will correspond to the proposed chronozones. The Bølling Chronozone thus comprises the Oldest Dryas and Bølling periods or pollen zone Ia and Ib of Iversen (1954), pollen zones DRI and BÖ of Nilsson (1961) and Brøndmyr Interstadial and preceding zones (I, II and III) of Faegri (1940) ...". Clearly then the chronozone boundaries have been partly positioned on the basis of inferred climate. Later (p.118) they state that these boundaries "seem to be climatically conditioned within southern Scandinavia".

As described above, Pennington (1975a) has suggested that the chronostratigraphic scheme devised for Norden can be extended to western Britain, but she also refers to climatic synchroneity between the two regions. Thus although a chronostratigraphic scheme has been proposed for NW Europe, the authors have looked beyond their definitions to see how they relate to probable climatic developments. Perhaps because of this Watts (this volume, p.3) suggests that chronostratigraphy "may be superfluous". The addition of specific names implies by itself that important distinctions are made between different time intervals and it is clear from many statements that these distinctions are climatic. Hence indirectly climatic units are advanced, and thus it would seem logical to attempt direct climatostratigraphic subdivision from the outset.

Another difficulty with chronostratigraphic classification of the Lateglacial is that it is in danger of leading to a suppression of one of the most potentially interesting aspects of Lateglacial history, *viz.* the time-transgressive nature of the changes. Thus because by definition chronozones have boundaries that are "isochronous surfaces" (Hedberg, 1976, p.67), and because environmental change is time-transgressive, it follows that when traced laterally chronozone boundaries will not be in accord with times of significant environmental change. We have already pointed out the possibility of this for the Scottish Lateglacial (Gray and Lowe, 1977) in relation to the chronozone boundaries of Mangerud *et al.* (1974). We feel that this is the major drawback to the application of chronostratigraphy to the Lateglacial. It does not appear to be easily adaptable to time-transgressive changes unless numerous regional chronostratigraphies are to be adopted with the attendant proliferation of terminology and potential confusion.

PROBLEMS WITH RADIOCARBON DATES, THEIR INTERPRETATION AND APPLICATION

In terms of absolute dating of Lateglacial stratigraphy, radiocarbon is the only technique applicable to suitable sediments or materials throughout the world. Advances have been made in varve chronology in recent years (*e.g.* Tauber, 1970; Vogel, 1970) but it remains applicable only to selected regions of the world (*e.g.* southern Sweden - see Berglund, 1979). It is possible that continuous dendrochronological records will eventually stretch back to the Lateglacial, but again this will be of major importance only in selected parts of the world, for the development of a valid tree-ring series requires unusual fossil records and is extremely time-consuming (see Pilcher, 1973; Fritts, 1976). While both varve and tree-ring dating may one day help to isolate and evaluate dating errors associated with Lateglacial radiocarbon dates, it is likely that for some time to come an absolute dating framework for the Lateglacial will rest almost exclusively on radiocarbon dates.

However, radiocarbon dating of the Lateglacial stratigraphy is fraught with difficulties. Many of the problems and possible error sources are well known, *e.g.* contamination by older or younger carbon and isotopic fractionation, and are ably summarised in the literature (*e.g.* Shotton, 1972; Olsson, 1974; West, 1977; Bowen, 1978; Berglund, 1979). Others have received less attention, *e.g.* the thickness of dated sediment slices (Gray and Lowe, 1977), graphite derived from bedrock (Sutherland, this volume), late melt-out of ice, and there is growing concern that hard-water error may be a ubiquitous problem in dating samples from freshly deglaciated terrain (Walker and Lowe, 1979a; Sutherland, this volume; Lowe and Walker, this volume) for "few rocks exist that do not contain carbon" (Sutherland, this volume, p.142).

To these difficulties should be added the overall perspective of the limitations of the dating technique and the time-span being considered, *i.e.* the resolution of the dating method. Let us consider the dating of the boundaries of the Younger Dryas (Loch Lomond Stadial) on the assumption that it lasted approximately 1,000 radiocarbon years. Most radiocarbon dates of sediments of Lateglacial or Flandrian age have standard deviations of about ±100 to ±250 radiocarbon years (*e.g.* Mangerud *et al.*, 1974, Table 2 ; Gray and Lowe, 1977, Table 1; Berglund, 1979, Table 2), though much larger ones (occasionally greater than ±450) have been reported. It must always be remembered that there is only a 68% probability that each date falls within the range of the standard deviation quoted (Harkness, 1975) and that to be statistically more rigorous two standard deviations should be used, giving a 95% probability that the "true" date falls within these limits, but still a 5% probability that it does not. Table 3 emphasises the resolution problems that arise if single dates of 11,000 BP and 10,000 BP are obtained with standard deviations of ±100 and ±250 radiocarbon years.

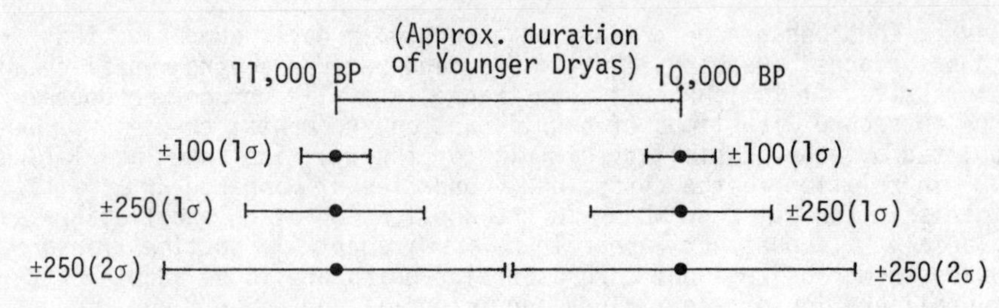

Table 3 Illustration of resolution problems in dating the Younger Dryas boundaries. For explanation see text.

In order to overcome a tendency to forget the standard deviations applied to individual dates, Sutherland (this volume, p.139) has suggested that dates expressed as a range covering one standard deviation about the mean "conveys the uncertainty associated with the dates more than does the standard method of expression." An even more radical measure would be for laboratories not to quote dates at all, but rather the C14 activity. It is this that is being measured, not calendar years, nor even radiocarbon 'years', and it is generally appreciated that the C14 activity of Lateglacial sediments can be derived from different sources (see above).

The uncertainties associated with individual radiocarbon dates can be reduced by accumulating several dates for the same boundary. This has been carried out in many studies, but problems may remain. Figure 1 is a plot of the radiocarbon dates (listed in Table 2 of Mangerud *et al.*, 1974) with two standard deviation age ranges, that have been used in defining chronozone boundaries in Norden. The problems with individual dates are illustrated by those at the upper boundary of the Younger Dryas Chronozone, for there are dates where the 95% probability limits do not overlap with each other or with the chronozone boundary. The careful accumulation of several dates can greatly reduce uncertainty, but in view of all the difficulties discussed above it is difficult to see how precise absolute dates can be affixed to stratigraphic boundaries. This is particularly true for the Older Dryas Chronozone which is supposed to have lasted only 200 years! The small number of dates for the boundaries of this chronozone and their overlapping tendency (Fig. 1) cast doubts on the validity of the boundary dates given. Radiocarbon dating, at current levels of precision, cannot be used to support them.

Mangerud *et al.* (1974) have emphasised the need for critical selection of dates. We would support this and propose that in all proposed schemes there should be a discussion of possible error sources in all dates selected as well as in all those rejected, together with descriptions of the type of material dated, the δC13 activities, and thickness of sediment slices dated (perhaps giving thickness relative to the total depth of a profile, or part of a profile, at individual sites, as well as absolute thickness). Table 1 in Lowe and Walker (this volume) may provide a good model. They claim that, even when dates for the close of the Loch Lomond Stadial are available from a number of sites, in view of the difficulties associated with radiocarbon dating it is not possible to state precisely when the close of the stadial occurred in Scotland, and that it is extremely difficult to make sound correlations between sites where limnic-terrestrial sediment samples have been employed as a dating medium.

This argument extends to dating Lateglacial boundaries in general. For example, Mangerud *et al.* (1974, Table 2) in deciding the absolute ages of their Lateglacial chronozone boundaries examined dates from "levels at, just below or above the dis-

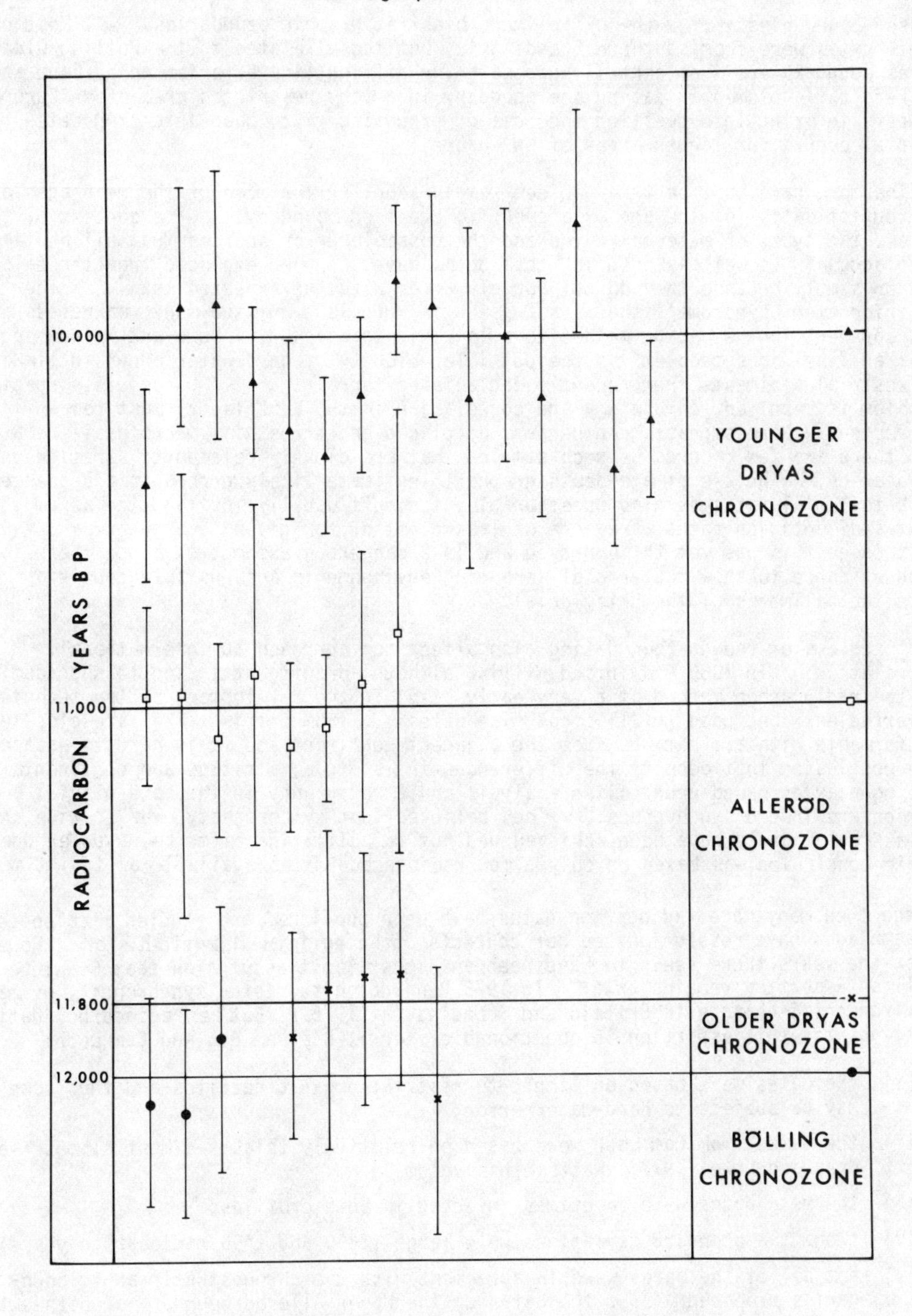

Fig. 1 Plot of the radiocarbon dates listed in Table 2 of Mangerud *et al.* (1974), with ±2 standard deviation ranges. Chronozone boundaries to which the dates relate are indicated by symbols.

cussed boundaries", *i.e.* the old climato-biostratigraphic boundaries. Most of their dates were from southern Scandinavia, but those related to the Bölling/Older Dryas boundary are from central Europe. The difficulties experienced by Lowe and Walker (this volume) in dating one boundary in a very restricted area of NW Europe appears to bring into question the idea of proposing fixed boundaries for Late-glacial events for larger areas of NW Europe.

Furthermore care must be taken in not simply adopting the mean of the mean ages of a group of dates for the age of a specific event or boundary. The context of the sites, the types of material dated and the possible error sources must all be taken into account, as well as ensuring that dates have not been excluded from the calculation simply because they do not roughly agree with the expected mean. The 'reinforcement syndrome' is always a potential problem, but too many weaknesses are now apparent in the dating method to allow this to develop. Lowe and Walker argue that an "insidious problem" is the possible ubiquity of hard-water error in limnic-terrestrial sediments from recently deglaciated terrain. Unless this special problem is resolved, chronology and correlation in the Lateglacial must remain imprecise. More emphasis can perhaps be placed on terrestrial macrofossil remains, but there are few records of such samples that are closely related to stratigraphic boundaries. The use of age-depth graphs to estimate likely errors in a sequence of Lateglacial dates is also questionable, for this usually involves the assumption that sedimentation rates were more or less constant for given time intervals. In most cases this has yet to be proved and is a dangerous assumption since there is much evidence to show that rapidly-changing environments and unstable conditions were the hallmarks of the Lateglacial.

The problems of radiocarbon dating also affect our approach to international correlation. In 1959 Godwin and Willis, although drawing attention to shortcomings in the radiocarbon method at a very early stage in the development of the technique, nevertheless concluded (p.211) that "The dates ... from British sites are closely conformable with ... datings from the adjacent continent and it is hard to escape the conclusion that despite the differences in latitude, altitude and continentality the boundaries based upon pollen analysis and stratigraphy in the Late-glacial Period are indeed synchronous." They believed that synchroneity over so wide an area "could hardly have been achieved had not the climatic shifts been sudden ones." Their conclusion was based on only a few radiocarbon dates available at that time.

Since then many more radiocarbon dates have been published, and studies of coleopteran assemblages have revolutionised our concepts of Lateglacial climatic history. However, over the years there seems to have been an almost tacit assumption that Godwin and Willis' assertion remains true. In 1975 Pennington suggested synchroneity of major environmental changes in Britain and Scandinavia (1975a). But her radiocarbon dating evidence for this assertion is questionable, for at Blelham Bog and Cam Loch:

- (i) the dates were based on limnic-terrestrial organic detritus and thus some may be subject to hard-water error;
- (ii) the dates from Cam Loch were based on relatively thick sediment slices (see Gray and Lowe, 1977; Watts, this volume);
- (iii) the mean dates were sometimes inverted in both profiles;
- (iv) two of the standard deviations were large (±450 and ±490 radiocarbon years);
- (v) not all of the dates were in agreement with the chronostratigraphic boundaries proposed (*e.g.* the dates on the Upper Alleröd boundary at both sites).

It is the contention of the authors, therefore, that it is not possible, with present information and available techniques, to date Lateglacial stratigraphic boundaries precisely or to decide whether or not many Lateglacial events were synchronous for different parts of NW Europe.

PROBLEMS WITH STRATOTYPE, TYPE SITE OR REFERENCE SECTION PROPOSALS

Although the establishment of stratotypes, type sites or reference sections has often resulted in the avoidance of both the ambiguity and lack of clarity that characterised some earlier work since the concept demands the most rigorous application of available techniques (*e.g.* Coope and Pennington, 1977), problems remain. For example there is the problem of what areal scale should be used - should we have stratotypes for the whole of NW Europe, or should there be separate stratotypes for parts of it and if so what size should these parts be? Is it valid to extend definitions based on one site to other regions when it has first to be established that similar developments occurred during the Lateglacial in different regions, and that synchronous changes took place? As Mangerud *et al.* (1974, p.111) point out, environmental changes may not be "recognisable outside limited areas".

The "Windermere Interstadial", which is now being applied in Britain (Coope and Pennington, 1977; Sissons, 1979c), illustrates this problem and introduces others. Firstly, it is not clear what type of stratigraphic unit is being defined, although the implication of the use of "Interstadial" and of the last column in their Table 1 headed "duration of interstadial climate" is that it is a climatostratigraphic one. Certainly G.F. Mitchell and R.G. West in the discussion that follows the proposal refer to a climatic boundary for the base of the Windermere Interstadial. But if it is a climatostratigraphic unit which of the two definitions of interstadial climate are we to accept? And how are we to use the term as a reference site for Britain?

For example, Pennington (1977a, p.247) states that "Between 13,000 and 11,000 there is in western Britain evidence for a woodland biozone, or palaeobotanical interstadial, equivalent to Bølling plus Allerød of continental stratigraphers ... and divided by a very minor regression of vegetation during Older Dryas time (*ca.* 12,000 - 11,800). The now closely C14-dated site at Low Wray Bay, Windermere, is suggested as a reference site for this interstadial". However, later she states that recently published data "... point to an absence of any Late Devensian woodland from much of western Britain, and to pronounced differences between pollen zones from one part of Britain to another. In addition, radiocarbon dating has shown that at sites in western Britain where significant expansion of woodland ... has been found,the zone of maximum development of woodland must have been non-synchronous ..." (p.248). The use of "Windermere Interstadial" for deposits from a site at Corstorphine, Edinburgh (Sissons, 1979c) exemplifies the sort of problem that arises. No woodland biozone is recognised in the original pollen diagram of Newey (1970). In Newey's own words "The herbaceous dominance characteristic of the Late-Weichselian zones appears to have been particularly marked at Corstorphine and even during the relative warmth of Zone II, tree-pollen frequencies are very low, particularly in the middle part of this zone, where they do not exceed 15%. Most of this pollen was from *Betula* and some of it was probably the result of long-distance transport" (p.1175). Further, no radiocarbon dates were reported from Corstorphine. Thus how can the biostratigraphic data from Corstorphine be equated with the data from Low Wray Bay? A precise correlation cannot be made.

There are other problems with the Low Wray Bay "reference section". Some of the mean radiocarbon date values are inverted in the series and some of the basal dates may be subject to hard-water error (Coope and Pennington, 1977, Table 1; Pennington, 1977a). The end product therefore is a "reference section" with two interpretations and an uncertain dating framework. This is hardly satisfactory. The purpose of formal stratotypes in that they should be clear and unambiguous, but the use of the term "Windermere Interstadial" would seem to us to be far from this.

It is important to stress that criticism is not levelled at the methods of analysis used at Low Wray Bay, nor the biostratigraphic subdivisions proposed. What is being emphasised is the need to recognise which of the two interpretations is being used and the problems with the radiocarbon dates at the reference site.

These arguments can be extended to procedures in general for the definition and terminology of the Lateglacial stratigraphy of NW Europe. With the current state of knowledge and apparent constraints of available techniques, particularly those of radiocarbon dating (see above), individual sites can only give limited spatial and uncertain temporal information. In view of this it can be argued that, at least for the present, definitions should take into consideration combined information from a number of sites. Even then boundary definitions may still be difficult (Lowe and Walker, this volume). It is relevant to note that in effect this was the procedure used by Mangerud *et al.* (1974 - see above) and that they refer (p.119) to southern Scandinavia as an "informal 'type' area" for their Lateglacial chronostratigraphy.

PROBLEMS WITH TERMINOLOGY

There are well-known definitional and terminological problems that exist in Lateglacial stratigraphic schemes and elsewhere in geological time. West (1977, p.259) suggests the use of the term "interglacial to describe a temperate period with a climatic optimum at least as warm as the Flandrian climatic optimum in the region concerned, and interstadial to describe a period which was either too short or too cold to permit the development of temperate deciduous forest of the interglacial type in the region concerned". Other attempts have been made to define how climatostratigraphic units or their equivalent should be defined, but, though many have been aware of the problems, fewer attempts have been made to suggest where the boundaries should be drawn. What degree of temperature change is required, or what value of temperature should be achieved to distinguish between 'interstadial' and 'stadial' conditions? It is partly because of these difficulties that differences exist in whether an Older Dryas stratigraphic unit should be established or where the base of the "Windermere Interstadial" should be placed. Since environmental changes are continuous, sharp stratigraphic boundaries are liable to give a misleading impression.

Another terminological problem concerns Watts' (this volume, p.21) interesting suggestion that "The Lateglacial has the character of an interglacial climatic/vegetational cycle that made a false start and had to begin again", and that the Younger Dryas in NW Europe "is merely a delaying factor in a little-changed succession in Central Europe". If this concept is correct, support is given to those who would extend the start of the Holocene back to 13,000 BP or earlier (*e.g.* Mercer, 1972), from its presently defined position at 10,000 BP. But the *status quo* can be supported by arguing that the beginning of the Lateglacial marks a "false end" to the last glaciation!

DISCUSSION OF THE SUGGESTED CLIMATOSTRATIGRAPHIC SCHEME

Summary of Problems

The suggested general climatostratigraphic scheme shown in Table 2 takes into account many of the problems and points raised in previous sections of this paper and summarised below.

(i) Some standard stratigraphic procedures are ill-suited to the Lateglacial. In the interests of conveying a simpler stratigraphic picture than presently exists for NW Europe we would suggest that a general climatostratigraphic scheme and terminology has much to commend it.

(ii) Since the boundaries of proposed chronostratigraphic schemes have been checked to determine whether or not they correspond with times of climatic change, it would seem more logical to set out to attempt a climatostratigraphic subdivision that can then be defined in terms of absolute chronology.

(iii) An Older Dryas climatic deterioration is not strongly represented throughout NW Europe. Thus we have constructed only two stratigraphic units for the Lateglacial. We feel that the terms 'Bölling', 'Older Dryas' and 'Alleröd' should be avoided for the present to avoid the more tenuous implications of this terminology.

(iv) Chronostratigraphy does not appear to be easily adaptable to time-transgressive changes in environment. Its boundaries are isochronous surfaces whereas our scheme has transitions partly to denote the possibility of time-transgression.

(v) Stratotype definitions may not be valid away from the type section, thus underlining the need for a flexible scheme such as ours.

(vi) Radiocarbon dating is fraught with difficulties, and its resolving power and error sources do not permit precise definition of chronozone boundaries. This is another reason why we have suggested transitional periods of time rather than 'boundaries'. Problems in radiocarbon dating also mean that uncertainty frequently exists in the dates from single sites, including stratotype sections. Schemes constructed on the basis of combinations of dated sites or combinations of evidence may thus be preferable (see below).

(vii) Since environmental changes do not occur at points in time, transitions to represent these changes are more realistic.

General Discussion

We envisage our climatostratigraphic scheme as a general one that could be applied throughout NW Europe. We would support the definitions of the American Code (1961) and of Mangerud *et al.* (1974) given earlier in this paper that climatic units should be based on some other type of stratigraphic unit such as biostratigraphy It may be necessary to use combinations of stratigraphic units in defining climatic units and/or to select those that are believed to be the most valid indicators of climatic changes. In this respect coleopteran evidence, though not free of difficulties appears to have considerable advantages over other terrestrial biota and it seems likely that, where available, coleopteran units will often provide the best stratigraphic units on which to base Lateglacial climatostratigraphy. Climate, however, is a complex variable and Coleoptera mainly indicate variations in one climatic parameter - mean July temperature. Temperature is nevertheless a very important aspect of climatic change and it is the one most likely to be reconstructed from Lateglacial sediments and hence the one on which climatostratigraphy will commonly be based.

As indicated earlier, one of the main criticisms of using climate in stratigraphic classification is that climate itself is not observed in the rocks, only the effects of climate. Thus the units are interpretive or inferred. Hedberg (1976, p.95), however, makes a similar observation about chronostratigraphy - "Although lithostratigraphic and biostratigraphic units are largely established and distinguished on the basis of observable physical features, chronostratigraphic units are identified on the basis of their time of formation - a more interpretive character." Similarly Vita-Finzi (1973, p.7) refers to "the 'inferential' basis of chronostratigraphy". This certainly applies to the type of chronozones defined by Mangerud *et al.* (1974) for the Lateglacial in Norden for they are not based on stratotypes but on combinations of evidence. Thus in terms of this major criticism of climatostratigraphy, the chronostratigraphic scheme of Mangerud *et al.* has little advantage over our suggested scheme.

Although perhaps not immediately apparent, there are more similarities between the scheme of Mangerud *et al.* and our suggested one than might be supposed. Firstly, it has already been suggested that the scheme of Mangerud *et al.*, although termed chronostratigraphic, is partly based on and closely allied to climate. Secondly, both sets of authors have preferred combinations of evidence rather than stratotypes in defining boundaries. Thirdly, had we used line boundaries instead of transitions our scheme would have been a chronostratigraphic one based on climatostratigraphic units (*cf.* the Mangerud *et al.* chronostratigraphic scheme based on "the old climato-biostratigraphic zones").

Although we suggest a climatostratigraphic scheme for NW Europe, we appreciate that workers in certain areas may find it desirable to use national or regional equivalent terms. Thus we see no inconsistency in the continued use of the term "Loch Lomond Stadial" in Britain or the "Nahanagan Stadial" in Ireland. We suggest however that these are quoted as, for example, "Loch Lomond (Younger Dryas) Stadial", but we suggest that for simplicity the number of these local equivalents be kept to a minimum across NW Europe.

A more important reason for the introduction of regional terminologies, however, would come if strong evidence were to evolve for time-transgressive climatic change across NW Europe during the Lateglacial. This might then enable us to define narrower transition zones for different regions, and although this might mean a proliferation of regional terminology, we would still have the general climatostratigraphy and its general terms to which this could easily be related.

In addition unequivocal evidence for a climatostratigraphic subdivision of the Lateglacial Interstadial may eventually unfold. Until such time, however, local stratigraphies should be given non-formal local zonation nomenclature only (*e.g.* subzones a, b, c). This follows the philosophy behind recent procedures in pollenzonation subdivisions (Birks, 1973) of the Lateglacial. Only when a firmer data bank is established should a more complicated climatostratigraphic scheme be established. We see this step as being a long and necessarily painstaking one.

Terms and Boundaries

Finally it is necessary to discuss the terminology and the positioning of the dashed lines shown in Table 2. We would emphasise that these lines are not rigidly fixed - they are dashed to indicate flexibility, for it may be necessary to widen or possible to reduce the transition zones if this appears appropriate in the light of future research.

The term "Lateglacial Interstadial"[1] has been used since it is a general term that can be applied to that part of the Lateglacial record that precedes the Younger Dryas. Until the introduction of the term "Windermere Interstadial" it had begun to gain general acceptance in Britain (*e.g.* Sissons, 1974b, 1976b; Coope, 1975, 1977; Walker, 1977).

---

[1]It might be argued that strictly the terms 'Late-Devensian lateglacial' or 'Late-Weichselian lateglacial' should be employed. However, we favour the use of 'Lateglacial' without a prefix to refer to the end of the last glacial stage and the use of prefixes only for lateglacial stages relating to earlier glaciations *e.g.* 'Anglian-lateglacial' or 'Saalian-lateglacial'.

The beginning of the Lateglacial Interstadial is difficult to define. Much of the evidence was reviewed by Gray and Lowe (1977, pp.176-78) and the transition shown in Table 2 mirrors our statement (p.178) that current evidence suggests the start of a marked thermal improvement "occurred between about 14,000 and 13,000 BP". The earlier date is in accord with the suggestion by G.F. Mitchell and R.G. West (in discussion following Coope and Pennington, 1977) that "14,000 BP is a position which has some significance for the evolution of the climate in the Late-Devensian", though this age is derived from the basal dates from Low Wray Bay (see above). Ruddiman and McIntyre (1973) found that polar water retreated from Britain around 13,500 BP, but the fact that the retreat was time-transgressive over NW Europe and the fact that the date is derived by interpolation based on assumed sedimentation rates, further emphasises the need for a transition zone. Coope (1977, p.313) concludes that in Britain there was a sharp temperature rise "before 13,000 years ago but later than 14,000 years ago". Berglund (1979, p.110) claims that in southern Sweden "... the deglaciation pattern and the vegetation/soil changes indicate an important climatic amelioration slightly after 13,000 BP ...", but constructs his lower, interstadial, climatostratigraphic boundary (which he shows as a shaded time band) at around 13,000 BP.

The close of the Lateglacial Interstadial is even more problematic. A fixed boundary for this (or its equivalent) has been placed by many workers at or about 11,000 BP (*e.g.* Mangerud *et al.*, 1974; Pennington, 1975a; Gray and Lowe, 1977; Berglund, 1979). On the other hand Sissons (1979c, p.202) points out that "Coleoptera indicate that a decline in temperature began before 12,000 BP, Cladocera imply a deteriorating climate perhaps from *c.* 11,800 BP, and some pollen evidence indicates irregularly declining temperatures from *c.* 11,800 BP". Coope (1977) advocates a sharp decline in temperature at about 11,000 BP that was more marked than earlier temperature cooling, but we have begun our transition zone at 12,000 BP to indicate that a cooling trend appears to have begun at about that time.

The term "Younger Dryas" is firmly established in the European literature, and we see no reason to replace the term with an alternative. The upper boundary of this stadial has not been satisfactorily dated. The Holocene Commission for INQUA fixed the lower Holocene boundary at 10,000 BP and Mitchell *et al.* (1973) recommended the same date for the beginning of the Flandrian in Britain. Similarly Mangerud *et al.* (1974) adopted this date for the upper boundary of the Younger Dryas Chronozone in Norden and an age of 10,000 BP has been suggested for a boundary-stratotype at Gothenburg, Sweden (Mörner, 1976). On the other hand there is evidence of climatic amelioration at a somewhat earlier date. Mangerud *et al.* (1974, p.120) pointed out that dates from southern Scandinavia on the biostratigraphic boundary related to the close of the Younger Dryas were slightly older than 10,000 BP, and Berglund (1979, p.110) refers to a "severe, long period of deterioration 11,000 to 10,200 B.P., corresponding to the Younger Dryas". In Britain evidence has recently been put forward to suggest the possibility of climatic improvement before 10,200 and perhaps as early as 10,400 BP (Lowe and Walker, 1976, this volume; Pennington, 1977b). Thus a provisional age-range for the transition from the Younger Dryas Stadial to the marked climatic improvement of the Flandrian Interglacial would be between about 10,500 and 10,000 BP.

It should be noted that some amendments have been made to the definitions of the two climatostratigraphic units given in a previous paper (Gray and Lowe, 1977) and repeated in the introduction to this book. Firstly, both our definitions began "The period between ..." but according to Hedberg (1976, pp.10-11) such units of geological time, being intangible, are not in themselves stratigraphic units. To indicate that we believe that climatostratigraphic units should be based on properties of sediment strata, we now propose that our definitions of climatostratigraphic units should begin "The body of rock (sediment) strata formed between ...". Secondly it will be noted from Table 2 that the term "Lateglacial Interstadial" is now suggested for NW Europe and its definition and duration have therefore been modified.

Provisional definitions for the climatostratigraphic units suggested for NW Europe are therefore as follows:

Lateglacial Interstadial: The body of rock (sediment) strata formed between the marked thermal improvement that occurred between about 14,000 and 13,000 BP and the thermal decline that took place between about 12,000 and 11,000 BP.

Younger Dryas Stadial: The body of rock (sediment) strata formed between the thermal decline that took place between about 12,000 and 11,000 BP and the marked thermal improvement that took place between about 10,500 and 10,000 BP.

Lateglacial: The body of rock (sediment) strata formed between the start of the Lateglacial Interstadial and the end of the Younger Dryas Stadial.

CONCLUSIONS

Above all our suggested scheme is intended to convey the message that there does seem to be considerable agreement on the Lateglacial climate of NW Europe. Without it we may soon be lost in a maze of alternative names, dates and details of local stratigraphies.

To many the scheme outlined above will seem unsatisfactory and they will point to its lack of precision and partial disregard of accepted stratigraphic procedures. However, we have argued above that the Lateglacial may be considered to be a special case, and that neither are some existing schemes and procedures entirely in line with the recommendations of the ISSC *Guide*.

We would argue that our suggested scheme encompasses the desirable qualities of flexibility, simplicity, and accuracy in reflecting the uncertainty of present knowledge, particularly in radiocarbon dating. While fundamental issues remain debatable, any strict stratigraphic framework for NW Europe introduces a false precision to our real knowledge of Lateglacial stratigraphy. Although we know more about the Lateglacial than any other unit of geological time, with the exception of the Flandrian, our knowledge is in many respects still limited and incomplete, and our stratigraphic framework should reflect this. Bowen (1978, p.197) appears to capture the spirit of this in discussing the correlation of Quaternary stratigraphy. He points out that the evidence in most regions is so problematic that "... the use of a floating chronology ... is desirable, wherein units are flexibly disposed temporally, and are readily amenable to changing emphasis due to new discoveries or interpretations."

Perhaps it is also worth emphasising that no stratigraphic scheme can be completely satisfactory. We have pointed out difficulties in other schemes but there are also problems with ours. One of these is how a scheme for NW Europe can be reconciled with other parts of Europe. This problem and all those introduced above will only be resolved by discussion and further research. The main reason for writing this paper was to stimulate discussion of these important points for like Mangerud and Berglund (1978, p.180) we feel that NW Europe is a natural geographical region for a common Quaternary stratigraphic nomenclature.

ACKNOWLEDGEMENTS

We are very grateful to Dr. H.J.B. Birks, Mr. J. Ince, Dr. P.D. Moore, Mr. J. Rose and Dr. M.J.C. Walker for their constructive comments on the first draft of this paper. The version published here owes much to them and to many others who have argued and discussed these topics with us in recent years. However, we take full responsibility for the opinions expressed here.

References

AARSETH, I. and MANGERUD, J. 1974. Younger Dryas end moraines between Hardangerfjorden and Sognefjorden, Western Norway. *Boreas*, 3, 3-22.

ADAMS, T.D. and HAYNES, J. 1965. Foraminifera in Holocene marsh cycles at Borth, Cardiganshire (Wales). *Palaeontology*, 8, 27-38.

AMERICAN COMMISSION ON STRATIGRAPHIC NOMENCLATURE. 1961. Code of Stratigraphic Nomenclature. *Bull. Am. Ass. Pet. Geol.*, 45, 646-65.

ANDERSEN, A-L.L. 1971. Foraminifera from the Older Yoldia Clay at Hirtshals. *Medd. Dansk Geol. Foren.*, 21, 159-84.

ANDERSEN, B.G. 1960. Sørlandet i Sen- og Postglacial tid. *Norges Geol. Unders.*, 210, 142 pp.

ANDERSEN, B.G. 1968. Glacial geology of Western Troms, North Norway. *Norges Geol. Unders.*, 256, 160 pp.

ANDERSEN, B.G. 1979. The deglaciation of Norway 15,000 - 10,000 BP. *Boreas*, 8, 79-87.

ANDERSEN, S.Th. 1961. Vegetation and its environment in Denmark in the Early Weichselian Glacial. *Danm. Geol. Unders.*, Ser. II, 75, 1-175.

ANUNDSEN, K. 1977. Radiocarbon datings and glacial striae from the inner part of Boknfjord area, South Norway. *Norsk Geogr. Tidsskr.*, 31, 41-54.

ANUNDSEN, K. 1978. Marine transgression in Younger Dryas in Norway. *Boreas*, 7, 49-60.

ARAYA, R. and HERVÉ, F. 1972. Patterned Gravel Beaches in the South Shetland Islands. *In* ADIE, R.J. (ed.) *Antarctic Geology and Geophysics*, 111-14, Oslo.

BALLANTYNE, C.K. and WAIN-HOBSON, T. 1979. The Loch Lomond Advance on the Island of Rhum. *Scott. J. Geol.* (In press).

BARTLEY, D.D. 1962. The stratigraphy and pollen analysis of lake deposits near Tadcaster, Yorkshire. *New Phytol.*, 61, 277-87.

de BEAULIEU, J.L. 1977. Contribution pollenanalytique à l'histoire tardiglaciaire et holocène de la végétation des Alpes meridionales françaises. Diss., Marseille.

BECK, R.B., FUNNELL, B.M. and LORD, A.R. 1972. Correlation of Lower Pleistocene Crag at depth in Suffolk. *Geol. Mag.*, 109, 137-39.

BEHRE, K.E. 1967. The Late Glacial and Early Postglacial History of Vegetation and Climate in Northwestern Germany. *Rev. Palaeobotan. Palynol.*, 4, 149-61.

BEHRE, K.E. 1978. Die Klimaschwankungen im europäischen Präboreal. *Petermanns Geographische Mitteilungen*, 2, 97-102.

BENSON, R.H. 1971. A new Cenozoic deep-sea genus *Abyssocythere* (Crustacea: Ostracoda: Trachyleberididae), with descriptions of five new species. *Smiths. Contrib. Palaeobiol.*, 7, 1-25.

BERGLUND, B.E. 1971. Late-glacial stratigraphy and chronology in south Sweden in the light of biostratigraphic studies on Mt. Kullen, Scania. *Geol. Fören. Stockh. Förh.*, 93, 11-45.

BERGLUND, B.E. 1976. The deglaciation of southern Sweden. Presentation of a research project and a tentative radiocarbon chronology. *Univ. of Lund, Dept. of Quat. Geol. Rep.*, 10, 67 pp.

BERGLUND, B.E. 1979. The deglaciation of southern Sweden 13,500 - 10,000 BP. *Boreas*, 8, 89-118.

BERTSCH, A. 1961. Untersuchungen zur spätglazialen Vegetationsgeschichte Südwestdeutschlands. *Flora*, 151, 243-80.

BEUG, H.J. 1957. Untersuchungen zur spätglazialen und frühpostglazialen Floren- und Vegetationsgeschichte einiger Mittelgebirge. *Flora*, 145, 167-211.

BEUG, H.J. 1964. Untersuchungen zur spät- und postglazialen Vegetationsgeschichte im Gardaseegebiet unter besonderer Berücksichtigung der mediterranen Arten. *Flora*, 154, 401-44.

BEUG. H.J. 1976. Die spätglaziale und frühpostglaziale Vegetationsgeschichte im Gebiet des ehemaligen Rosenheimer Sees (Oberbayern). *Bot. Jahrb. Syst.*, 95, 373-400.

BIRD, J.B. 1967. *The Physiography of Arctic Canada.* Baltimore.

BIRKS, H.H. 1970. Studies in the vegetational history of Scotland I. A pollen diagram from Abernethy Forest, Inverness-shire. *J. Ecol.*, 58, 827-46.

BIRKS, H.H. and MATHEWES, R.W. 1978. Studies in the vegetational history of Scotland V. Late Devensian and early Flandrian pollen and macrofossil stratigraphy at Abernethy Forest, Inverness-shire. *New Phytol.*, 80, 455-84.

BIRKS, H.J.B. 1968. The identification of *Betula nana* pollen. *New Phytol.*, 67, 304-14.

BIRKS, H.J.B. 1973. *Past and Present Vegetation of the Isle of Skye: a palaeoecological study*. Cambridge.

BIRKS, H.J.B. and DEACON, J. 1973. A numerical analysis of the past and present flora of the British Isles. *New Phytol.*, 72, 877-902.

BISHOP, W.W. and COOPE, G.R. 1977. Stratigraphical and faunal evidence for Lateglacial and early Flandrian environments in south-west Scotland. *In* GRAY, J.M. and LOWE, J.J. (eds.) *Studies in the Scottish Lateglacial Environment*, 61-88, Oxford.

BÖCHER, T.W. 1952. Contributions to the flora and plant geography of West Greenland III. Vascular plants collected or observed during the botanical expedition to West Greenland, 1946. *Medd. om Grønl.*, 147 (9), 1-85.

BÖCHER, T.W. 1954. Oceanic and continental vegetational complexes in southwest Greenland. *Medd. om Grønl.*, 148 (1), 1-336.

BÖCHER, T.W. 1963. Phytogeography of Middle West Greenland. *Medd. om Grønl.*, 148 (3), 1-289.

BONATTI, E. 1966. North Mediterranean climate during the last Würm glaciation. *Nature*, 209, 984-5.

BONATTI, E. 1970. Pollen sequence in the Lake Sediments in Ianula. *In* HUTCHINSON, G.E. (ed.) *An account of the history and development of the Lago di Monterosi, Latium, Italy*, (Trans. Amer. Phil. Soc., 60), 26-31.

BORTENSCHLAGER, I. 1976. Beiträge zur Vegetationsgeschichte Tirol II, Kufstein-Kitzbuhel-Pass, Thurn. *Ber. Nat-med. Ver. Innsbrück*, 63, 105-37.

BOTTEMA, S. 1967. A Late Quaternary pollen diagram from Ioannina, northwestern Greece. *Proc. Prehist. Soc.*, n.s. 33, 26-9.

BOTTEMA, S. 1978. The Late Glacial in the Eastern Mediterranean and the Near East. *In* BRICE, W.C. (ed.) *The Environmental History of the Near and Middle East Since the Last Ice Age*, 15-28, London.

BOULTON, G.S., JONES, A.S., CLAYTON, K.M. and KENNING, M.J. 1977. A British Ice Sheet Model and patterns of glacial erosion and deposition in Britain. *In* SHOTTON, F.W. (ed.) *British Quaternary Studies*, 231-46, Oxford.

BOWEN, D.Q. 1978. *Quaternary Geology: a stratigraphic framework for multidisciplinary work*. Oxford.

BRADLEY, W.C. 1958. Submarine abrasion and wave-cut platforms. *Bull. Geol. Soc. Amer.*, 69, 967-74.

BRADY, G.S., CROSSKEY, H.W. and ROBERTSON, D. 1874. The Post-Tertiary Entomostraca of Scotland. *Mon. Pal. Soc. London*, 229 pp.

BRAUN-BLANQUET, J. 1965. *Plant Sociology: the Study of Plant Communities*. New York.

BROTZEN, F. 1951. Bidrag till de Svenska marina kvartaravlagri-ingarnas stratigrafi. *Geol. Fören. Stockh. Förh.*, 73, 57-68.

BROWN, A.P. 1971. The *Empetrum* pollen record as a climatic indicator in the Late Weichselian and Early Flandrian of the British Isles. *New Phytol.*, 70, 841-9.

BROWN, A.P. 1977. Late Devensian and Flandrian vegetational history of Bodmin Moor, Cornwall. *Phil. Trans. R. Soc. Lond. B*, 276, 251-320.

BROWNE, M.A.E., HARKNESS, D.D., PEACOCK, J.D. and WARD, R.G. 1977. The date of deglaciation of the Paisley-Renfrew area. *Scott. J. Geol.*, 13, 301-3.

CALKIN, P.E. and NICHOLS, R.L. 1972. Quaternary studies in Antarctica. *In* ADIE, R.J. (ed.) *Antarctic Geology and Geophysics*, 625-43, Oslo.

CASPARIE, W.A. and van ZEIST, W. 1960. A Late-glacial lake deposit near Waskemeer (Prov. of Friesland). *Acta. Bot. Neerl.*, 9, 191-6.

CASTON, V.N.D. 1977. Quaternary deposits of the Central North Sea, 1 & 2. *Rep. Inst. Geol. Sci.*, No. 77/11, 1-22.

CHANDA, S. 1965. The history of vegetation of Bröndmyra. A late-glacial and Early Post-glacial deposit in Jaeren, South Norway. *Acta Univ. Bergensis, Ser. Mat.-Rer. Nat.* 1965 (1), 1-17.

CHESHER, J.A., DEEGAN, C.E., ARDUS, D.A., BINNS, P.E. and FANNIN, N.G.T. 1972. IGS marine drilling with m.v. *Whitethorn* in Scottish waters, 1970-71. *Rep. Inst. Geol. Sci.* No. 72/10, 1-25.

CLIMAP Team. 1976. The surface of Ice-Age Earth. *Science*, 191, 1131-37.

CLINE, R.M. and HAYS, J.D. (eds.) 1976. *Investigation of Late Quaternary Palaeoceanography and Palaeoclimatology* (Geol. Soc. Amer. Memoir 145), 464 pp.

COACHMAN, L.K., HEMMINGSEN, E., SCHOLANDER, P.F., ENNS, T. and de VRIES, H. 1958. Gases in Glaciers. *Science*, 127, 1288-89.

COLHOUN, E. and SYNGE, F.M. 1979. The cirque moraines at Lough Nahanagan, County Wicklow, Ireland. *Proc. R. Irish Acad.*, (in press).

CONOLLY, A.P. and DAHL, E. 1970. Maximum summer temperature in relation to the modern and Quaternary distributions of certain arctic-montane species in the British Isles. *In* WALKER, D. and WEST, R.G. (eds.). *Studies in the Vegetational History of the British Isles*, 159-223, Cambridge.

COOPE, G.R. 1968. An insect fauna from mid-Weichselian deposits at Brandon, Warwickshire. *Phil. Trans. R. Soc. Lond. B*, 254, 425-56.

COOPE, G.R. 1970a. Interpretations of Quaternary insect fossils. *Ann. Rev. Ent.*, 15, 97-120.

COOPE, G.R. 1970b. Climatic interpretations of Late Weichselian coleoptera from the British Isles. *Revue Geogr. phys. Géol. dyn.*, 12, 149-55.

COOPE, G.R. 1975. Climatic fluctuations in northwest Europe since the last interglacial, indicated by fossil assemblages of Coleoptera. *In* WRIGHT, A.E. and MOSELEY, F. (eds.) *Ice Ages: Ancient and Modern*, 153-68, Liverpool.

COOPE, G.R. 1977. Fossil coleopteran assemblages as sensitive indicators of climatic changes during the Devensian (Last) cold stage. *Phil. Trans.*

R. Soc. Lond. B, 280, 313-37.

COOPE, G.R. and BROPHY, J.A. 1972. Late-glacial environmental changes indicated by a coleopteran succession from North Wales. *Boreas*, 1, 97-142.

COOPE, G.R. and OSBORNE, P.J. 1967. Report on the coleopterous fauna of the Roman well at Barnsley Park, Gloucestershire. *Trans. Bristol & Gloucs. Archaeol. Soc.*, 86, 84-7.

COOPE, G.R. and PENNINGTON, W. 1977. The Windermere Interstadial of the Late Devensian. *Phil. Trans. R. Soc. Lond. B.* 280, 337-39.

CRAIG, A.J. 1978. Pollen percentage and influx analyses in South-east Ireland. A contribution to the ecological history of the Late-glacial period. *J. Ecol.*, 66, 297-324.

CULVER, S.J. and BANNER, F.T. 1978. Foraminiferal assemblages as Flandrian palaeoenvironmental indicators. *Palaeogeogr. Palaeoclimatol. Palaeoecol.*, 24, 53-72.

CUSHING, E.J. 1967. Late-Wisconsin pollen stratigraphy and the glacial sequence in Minnesota. *In* CUSHING, E.J. and WRIGHT, H.E. (eds.) *Quaternary Palaeoecology*, 59-88, New Haven.

DAHL, E. 1956. Rondane: mountain vegetation in south Norway and its relation to the environment. *Skr. norske Videnskaps-Akad. I. Mat.-Nat.*, 3.

DAVIES, J.L. 1972. *Geographical Variations in Coastal Development.* Edinburgh.

DAVIS, M.B. 1969. Climatic changes in southern Connecticut recorded by pollen deposition at Rogers Lake. *Ecology*, 50, 409-22.

DAVIS, M.B. 1976. Pleistocene biogeography of temperate deciduous forests. *Geo-science and Man*, 13, 13-26.

DAVIS, M.B. 1978. Climatic interpretation of pollen in Quaternary sediments. *In* WALKER, D. and GUPPY, J.C. (eds.) *Biology and Quaternary Environments* Canberra.

DAWSON, A.G. 1977. A fossil lobate rock glacier in Jura. *Scott. J. Geol.*, 13, 37-42.

DAWSON, A.G. 1979. Raised shorelines of Jura, Scarba and NE Islay. Unpub. PhD thesis, Univ. of Edinburgh.

DAY, R.T. 1978. The autecology of *Diapensia lapponica* L. in Newfoundland. Unpub. BSc Diss., Dept. Biol., Memorial Univ. Newfoundland.

DEEVEY, E.S., GROSS, M.S., HUTCHINSON, G.E. and KRAYBILL, H.L. 1954. The natural C14 contents of materials from hard-water lakes. *Proc. U.S. Nat. Acad. Sci.*, 40, 285-88.

DICKSON, C.A., DICKSON, J.H. and MITCHELL, G.F. 1970. The Late-Weichselian flora of the Isle of Man. *Phil. Trans. R. Soc. Lond. B*, 258, 31-79.

DIETZ, C., GRAHLE, H.O. and MÜLLER, H. 1958. Ein spätglaziales Kalkmudde-Vorkommen im Seck-Bruch bei Hannover. *Geol. Jahrb.*, 76, 67-102.

DIONNE, J.C. 1973. Distinction entre stries glacielles et stries glaciares.

Rev. de Geogr. de Montreal, 27, 185-90.

DONNER, J.J. 1963. The Late- and Post-glacial raised beaches in Scotland. *Ann. Acad. Scient. Fenn., Ser. A, III*, 68, 1-13.

DONNER, J.J. 1978. The dating of the levels of the Baltic Ice Lake and the Salpausselkä moraines in South Finland. *Commentat. Physico-Math.*, 48, 11-38.

DONNER, J.J. and JUNGNER, H. 1973. The effect of re-deposited organic material on radiocarbon measurements of clay samples from Somero, southwestern Finland. *Geol. Fören. Stockh. Förh.*, 95, 267-8.

DONNER, J.J. and JUNGNER, H. 1974. Errors in the radiocarbon dating of deposits in Finland from the time of deglaciation. *Bull. Geol. Soc. Finland.* 46, 139-44.

DONNER, J.J., JUNGNER, H. and VASARI, Y. 1971. The hard-water effect on radiocarbon measurements of samples from Säynäjälampi, north west Finland. *Commentat. Physico-Math.*, 41, 307-10.

EASTWOOD, T., DIXON, E.E.L., HOLLINGWORTH, S.E. and SMITH, B. 1931. Geology of the Whitehaven and Workington district. *Mem. Geol. Surv. Engl. and Wales.*

EDEN, R.A., HOLMES, R. and FANNIN, N.G.T. 1978. Quaternary deposits of the central North Sea, 6. *Rep. Inst. Geol. Sci.* No. 77/15, 1-18.

EICHER, U. and SIEGENTHALER, U. 1976. Palynological and oxygen isotope investigations on Late Glacial sediment cores from Swiss lakes. *Boreas*, 5, 109-17.

EMILIANI, C. 1966. Palaeotemperature analysis of the Caribbean cores P.6304-8 and P.6304-9 and a generalised temperature curve for the last 425,000 years. *J. Geol.*, 74, 109-26.

FAEGRI, K. 1940. Quartärgeologische Untersuchungen im westlichen Norwegen. II. Zur spätquartären Geschichte Jaerens. *Bergens Mus. Arbok 1939-40. Naturv. rekke Nr 7*, 1-201.

FAEGRI, K. and IVERSEN, J. 1964. *Textbook of Pollen Analysis.* (second edition). Copenhagen.

FAEGRI, K. and IVERSEN, J. 1975. *Textbook of Pollen Analysis* (third edition). Oxford.

FEYLING-HANSSEN, R.W. 1964. Foraminifera in Late Quaternary deposits from the Oslofjord area. *Norges geol. Unders.*, 225, 383 pp.

FEYLING-HANSSEN, R.W. 1971. Weichselian interstadial foraminifera from the Sandnes-Jaeren area. *Medd. Dansk Geol. Foren*, 21, 72-116.

FEYLING-HANSSEN, R.W. 1972. The Pleistocene/Holocene boundary in marine deposits from the Oslofjord area. *Boreas*, 1, 241-46.

FEYLING-HANSSEN, R.W., JØRGENSEN, J.A., KNUDSEN, K.L. and ANDERSEN, A.-L.L. 1971. Late Quaternary foraminifera from Vendsyssel, Denmark and Sandnes, Norway. *Bull. Geol. Soc. Denmark*, 21, 67-317.

FIRBAS, F. 1950. The Late-glacial vegetation of Central Europe. *New Phytol.*, 49, 163-73.

FIRBAS, F., MÜLLER, H. and MÜNNICH, K.D. 1955. Das wahrscheinliche Alter der Späteiszeitlichen 'Bölling' Schwankung. *Die Naturwissenschaften*, 18, 509-10.

FISHER, M.J., FUNNELL, B.M. and WEST, R.G. 1969. Foraminifera and pollen from a marine interglacial deposit in the Western North Sea. *Proc. Yorks. geol. Soc.*, 37, 311-20.

FLORSCHÜTZ, F., MEMÉNDES AMOR, J. and WIJMSTRA, T.A. 1971. Palynology of a thick Quaternary succession in Southern Spain. *Palaeogeogr. Palaeoclimatol. Palaeoecol.*, 10, 233-64.

FORBES, E. 1846. On the connection between the distribution of the existing fauna and flora of the British Isles and the geological changes which have affected their area, especially during the epoch of the Northern Drift. *Mem. Geol. Surv. Gt. Br.*, vol. I.

FORD, D.C., FULLER, P.G. and DRAKE, J.J. 1970. Calcite precipitates at the soles of temperate glaciers. *Nature*, 226, 441-42.

FORD, E.B. 1976. *Genetics and Adaptation* (Edward Arnold, Studies in Biology No. 69). London.

FRANK, A.H.E. 1969. Pollen stratigraphy of the Lake of Vico (central Italy). *Palaeogeogr. Palaeoclimatol. Palaeoecol.*, 6, 67-85.

FREDÉN, C. 1978. Vänerområdets utveckling under de senaste 12,000 aren. *In* HÅKANSON, L., FREDÉN, C., LINDH, A., ROGNE, B. and STIGH, J. (eds.) *Vänerns morfometri och morfologi - en sjømorfometrisk handbok*, 70-79, Statens Naturvärdsverk, Solna.

FREDSKILD, B. 1975. A Late-glacial and early Post-glacial concentration diagram from Langeland, Denmark. *Geol. Fören. Stockh. Förh.*, 97, 151-61.

FRENCH, H.M. 1976. *The Periglacial Environment*. London.

FRITTS, H.C. 1976. *Tree Rings and Climate*. London.

FRITZ, A. 1972. Das Spätglazial in Kärnten. *Ber. Deutsch Bot. Ges.*, 85, 93-99.

FUNNELL, B.M. 1961. The Palaeogene and early Pleistocene of Norfolk. *Trans. Norfolk Norwich Nat. Soc.*, 19, 34-64.

FUNNELL, B.M. 1972. The history of the North Sea. *Bull. geol. Soc. Norfolk*, 21, 2-10.

GODWIN, H. 1975. *The History of the British Flora*. Cambridge.

GODWIN, H. and WILLIS, E.H. 1959. Radiocarbon dating of the Late-glacial period in Britain. *Proc. R. Soc. Lond. B*, 150, 199-215.

GOLDTHWAITE, R.P., LOEWE, F., UGOLINI, F.C., DECKER, H.F., DELONG, D.M., TRAUTMAN, M.B., GOOD, E.E., MERRELL, T.R. and RUDOLPH, E.D. 1966. Soil development and ecological succession in a deglaciated area of Muir Inlet, southeast Alaska. *Ohio State Univ. Inst. Polar Studies Rep.*, 20.

GORDON, A.D. and BIRKS, H.J.B. 1972a. Numerical methods in Quaternary palaeoecology I. Zonation of pollen diagrams. *New Phytol.*, 71, 961-79.

GORDON, A.D. and BIRKS, H.J.B. 1972b. Numerical methods in Quaternary palaeoecology II. Comparison of pollen diagrams. *New Phytol.*, 73, 221-49.

GRAY, J.M. 1974. The Main Rock Platform of the Firth of Lorn, western Scotland. *Trans. Inst. Brit. Geogr.*, 61, 81-99.

GRAY, J.M. 1978. Low-level shore platforms in the south-west Scottish Highlands: altitude, age and correlation. *Trans. Inst. Brit. Geogr., New Series*, 3, 151-64.

GRAY, J.M. and BROOKS, C.L. 1972. The Loch Lomond Readvance moraines of Mull and Menteith. *Scott. J. Geol.*, 8, 95-103.

GRAY, J.M. and LOWE, J.J. 1977. The Scottish Lateglacial Environment: a synthesis. *In* GRAY, J.M. and LOWE, J.J. (eds.) *Studies in the Scottish Lateglacial Environment*, 163-81, Oxford.

GREENSMITH, J.T. and TUCKER, E.V. 1971. Overconsolidation in some fine-grained sediments; its nature, genesis and value in interpreting the history of certain English Quaternary deposits. *Geol. en Mijnb.*, 50, 743-48.

GREGORY, D. and HARLAND, R. 1978. The Late Quaternary climatostratigraphy of IGS Borehole SLN 75/33 and its application to the palaeoceanography of the north-central North Sea. *Scott. J. Geol.*, 14, 147-55.

GRØNLIE, O.T. 1924. Contributions to the Quaternary Geology of Novaya Zemlya. *In* HOLTEDAHL, O. (ed.) *Report of the Scientific Results of the Norwegian Expedition to Nova Zemlya, 1921.*

GRÜGER, E. 1977. Pollenanalytische Untersuchung zur würmzeitlichen Vegetationsgeschichte von Kalabrien (Süditalien). *Flora*, 166, 475-89.

GRÜGER, J. 1968. Untersuchungen zur spätglazialen und frühpostglazialen Vegetationsentwicklung der Südalpen im Umkreis des Gardasees. *Bot. Jahrb. Syst.*, 88, 163-99.

GUILCHER, A. 1958. *Coastal and Submarine Morphology*. London.

GULLIKSEN, S., NYDAL, R. and SKOGSETH, F. 1978. Trondheim natural radiocarbon measurements VIII. *Radiocarbon*, 20, 105-33.

HALLET, B. 1976. Deposits formed by subglacial precipitation of $CaCO_3$. *Geol. Soc. Amer. Bull.*, 87, 1003-15.

HALLET, B., LORRAIN, R. and SOUCHEZ, R. 1978. The composition of basal ice from a glacier sliding over limestone. *Geol. Soc. Amer. Bull.*, 89, 314-20.

van der HAMMEN, T. 1951. Late-glacial flora and periglacial phenomena in the Netherlands. *Leidse Geol. Meded.*, 17, 71-183.

van der HAMMEN, T. and VOGEL, J.C. 1966. The Susacá-Interstadial and the subdivision of the Late-glacial. *Geol. en Mijnb.*, 45, 33-35.

HANDBOOK OF GEOCHEMISTRY. 1966. Springer-Verlag, Berlin.

HARKNESS, D.D. 1975. The role of the archaeologist in C-14 age measurement. *In*

WATKINS, T. (ed.) *Radiocarbon: Calibration and Prehistory*, 128-35, Edinburgh.

HARLAND, R., GREGORY, D.M., HUGHES, M.J. and WILKINSON, I.P. 1978. A late Quaternary bio- and climatostratigraphy for marine sediments in the north-central part of the North Sea. *Boreas*, 7, 91-6.

HARPER, J.L. 1969. The role of predation in vegetational diversity. *Brookhaven Symposia in Biol.*, 22, 48-62.

HAYNES, J. and DOBSON, M. 1969. Physiography, foraminifera and sedimentation in the Dovey Estuary (Wales). *Geol. J.*, 6, 217-56.

HAYNES, J.R., KITELEY, R.J., WHATLEY, R.C. and WILKS, P.J. 1977. Microfaunas, microfloras and the environmental stratigraphy of the Late Glacial and Holocene in Cardigan Bay. *Geol. J.*, 12, 129-58.

HAZEL, J.E. 1967. Classification and distribution of the Recent Hemicytheridae and Trachyleberididae (Ostracoda) off Northeastern North America. *U.S. Geol. Surv. Prof. Paper*, 564, 1-49.

HAZEL, J.E. 1970. Ostracode zoogeography in the Southern Nova Scotian and Northern Virginian faunal provinces. *U.S. Geol. Surv. Prof. Paper*, 529-E, 1-21.

HEDBERG, H.D. (ed.) 1970. Preliminary Report on Stratotypes. 39 pp. Int. Subcomm. Stratigraph. Class., Report 4, 24th Int. Geol. Congress, Montreal.

HEDBERG, H.D. (ed.) 1976. *International Stratigraphic Guide*. New York.

HESSLAND, I. 1943. Marine Schalenablagerungen Nord-Bohuslans. *Bull. geol. Inst. Uppsala*, 31, 348 pp.

HILLS, E.S. 1972. Shore platforms and wave ramps. *Geol. Mag.*, 109, 81-92.

HINXMAN, L.W. 1901. The River Spey. *Scott. Geogr. Mag.*, 17, 185-93.

HOBBIE, J.E. 1961. Summer temperatures in Lake Schrader, Alaska. *Limnol. Oceanogr.*, 6, 326-29.

HOEL, A. 1909. Geologiske iagttagelser på Spitsbergen ekspeditionerne 1906 og 1907. *Norsk geol. Tidsskr.*, 1, 1-28.

HOLMES, R. 1977. Quaternary deposits of the central North Sea, 5. The Quaternary geology of the UK sector of the North Sea between 56° and 58°N. *Rep. Inst. Geol. Sci.*, 77/14, 50 pp.

HOLTEDAHL, O. (ed.) 1929. On the geology and physiography of some Antarctic and sub-Antarctic islands with notes on the character and origin of fjords and strandflats of some Northern lands. *In Scientific Results of the Norwegian Antarctic Expeditions, 1927-8*, 172 pp.

HOLTEDAHL, O. 1960. Geology of Norway. *Norges Geol. Unders.*, 208, 507-30.

HUGHES, M.J., GREGORY, D.M., HARLAND, R. and WILKINSON, I.P. 1977. Late Quaternary foraminifera and dinoflagellate cysts from boreholes in the UK sector of the North Sea between 56° and 58°N. Appendix I. *In* HOLMES, R. (*op. cit.*), 36-46.

HUTCHINSON, G.E. 1957. *A Treatise on Limnology, Vol. I.* New York.

HYVÄRINEN, H. 1978. Use and definition of the term Flandrian. *Boreas*, 7, 182.

IVERSEN, J. 1954. The lateglacial flora of Denmark and its relation to climate and soil. *Danm. Geol. Unders.*, Ser. II, 80, 87-119.

JAHN, A. 1961. Quantitative analysis of some periglacial processes in Spitsbergen. *Uniwersytet Wroclawski Im Boleslawa Bieruta, Zeszyty Nauk. Nauki Przyrodnicze,* II, Warsaw.

JALUT, G. 1973. Analyse Pollinique de la Toubière de la Moulinasse: versant nord oriental des Pyrénées. *Pollen Spores*, 15, 472-509.

JAMIESON, T.F. 1865. On the history of the last geological changes in Scotland. *Q. J. geol. Soc. Lond.*, 21, 161-203.

JAMIESON, T.F. 1892. Supplementary remarks on Glen Roy. *Q. J. geol. Soc. Lond.*, 48, 5-28.

JARÁI-KOMLÓDI, M. 1970. Beiträge zum Spätglazial in Ungarn. *In Probleme der Weichsel-spätglazialen Vegetationsentwicklung in Mittel- und Nordeuropa*, 124-38, Frankfurt/Oder.

JESSEN, K. 1938. Some west Baltic pollen diagrams. *Quartär*, 1, 124-39.

JOACHIM, M.J. 1978. Late-glacial coleopteran assemblages from the west coast of the Isle of Man. Unpub. PhD Thesis, Univ. of Birmingham.

KARLEN, W. and DENTON, G.H. 1976. Holocene glacial variations in Sarek National Park, Northern Sweden. *Boreas*, 5, 25-56.

KARROW, P.F. and ANDERSON, T.W. 1975. Palynological study of lake sediment profiles from southwestern New Brunswick: discussion. *Can. J. Earth Sci.*, 12, 1808-12.

KING, C.A.M. 1972. *Beaches and Coasts.* London.

KIRK, W. and GODWIN, H. 1963. A late-glacial site at Loch Droma, Ross and Comarty. *Trans. R. Soc. Edinb.*, 65, 225-49.

KJEMPERUD, A. 1978. Strandforskyvning på Frosta, Nord-Trøndelag, belyst ved hjelp av palaeøkologiske metoder. Unpub. Thesis, Univ. of Trondheim.

KLOET, G.S. and HINCKS, W.D. 1945. *A check-list of British insects.* (Kloet and Hincks) Stockport.

KNUDSEN, K.L. 1971. Late Quaternary foraminifera from Vendsyssel, Denmark and Sandnes, Norway. *Medd. Dansk Geol. Foren.*, 21, 185-291.

KNUDSEN, K.L. 1978. Middle and Late Weichselian marine deposits at Nørre Lyngby, northern Jutland, Denmark, and their foraminiferal faunas. *Danm. geol. Unders.*, II, 112, 1-44.

KOMAR, P.D. 1976. *Beach Processes and Sedimentation.* New Jersey.

KRANCK, E. 1950. On the geology of the East Coast of Hudson Bay and James Bay. *Acta Geogr.*, 11, *Soc. Geogr. Fenn.*, Helsinki.

KROG, H. 1954. Pollen analytical investigation of a C14-dated Alleröd-section from Ruds-Vedby. *Danm. geol. Unders.*, Ser. II, 80, 120-39.

KROG, H. and TAUBER, H. 1974. C14 chronology of Late and Post-glacial marine deposits in North Jutland. *Danm. Geol. Unders.*, Årbog, 1973, 93-105.

KÜTTEL, M. 1974. Zum alpinen spät- und frühen Postglazial. Das Profil Obergurbs (1910 m) im Diemtigtal, Berner Oberland, Schweiz. *Z. Gletschcerk. u. Glazialgeol.*, 10, 207-16.

KÜTTEL, M. 1977. Pollenanalytische und geochronologische Untersuchungen zur Piottino-Schwankung (Jüngere Dryas). *Boreas*, 6, 259-74.

LAEVASTU, T. 1963. Serial atlas of the marine environment; Folio 4, surface water types of the North Sea and their characteristics. *Amer. Geog. Soc.* New York.

LANG, G. 1952. Zur späteiszeitlichen vegetations- und Florengeschichte Südwestdeutschlands. *Flora*, 139, 243-94.

LANG, G. and TRAUTMANN, W. 1961. Zur spät- und nacheiszeitlichen Vegetationsgeschichte der Auvergne (Französisches Zentralmassiv). *Flora*, 150, 11-42.

LASCA, N.P. 1969. Moraines in the Hemnefjord area, western Norway. *Norsk Geogr. Tidsskr.*, 24, 121-48.

LEE, G.F. 1970. Factors affecting the transfer of materials between water and sediments. *Univ. Wisconsin Water Resources Center, Eutrophication Information Programme.*

LEES, B.J. 1975. Foraminiferida from Holocene sediments in Start Bay. *Q. J. geol. Soc. Lond.*, 131, 37-49.

LESEMANN, B. 1969. Pollenanalytische Untersuchungen zur Vegetationsgeschichte des Hannoverschen Wendlandes. *Flora*, 158, 480-519.

LIVINGSTONE, D.A., BRYAN, K., Jr., and LEAHY, R.G. 1958. Effects of an arctic environment on the origin and development of freshwater lakes. *Limnol. Oceanogr.*, 3, 192-214.

LØFALDLI, M. 1973. Foraminiferal biostratigraphy of late Quaternary deposits from the Frigg Field and Booster Station. *Norges Teknisk - Naturvitenskapelige Forskningsrad, Continental Shelf Division, Rep.* 18, 1-82.

LORD, A.R. and ROBINSON, J.E. 1978. Marine Ostracoda from the Quaternary Nar Valley Clay, west Norfolk. *Bull. geol. Soc. Norfolk*, 30, 113-18.

LOWE, J.J. 1977. Pollen analysis and radiocarbon dating of Lateglacial and early Flandrian deposits in southern Perthshire. Unpub. PhD Thesis, Univ. of Edinburgh.

LOWE, J.J. 1978. Radiocarbon-dated Lateglacial and early Flandrian pollen profiles from the Teith Valley, Perthshire, Scotland. *Pollen Spores*, 20, 367-97.

LOWE, J.J. and WALKER, M.J.C. 1976. Radiocarbon dates and the deglaciation of Rannoch Moor, Scotland. *Nature*, 246, 632-33.

LOWE, J.J. and WALKER, M.J.C. 1977. The reconstruction of the Lateglacial environment in the southern and eastern Grampian Highlands. *In* GRAY, J.M. and LOWE, J.J. (eds.) *Studies in the Scottish Lateglacial Environment*, 101-18, Oxford.

LUDLAM, S.D. 1976. Laminated sediments in holomictic Berkshire lakes. *Limnol. Oceanogr.*, 21, 743-46.

McANDREWS, J.H. 1967. Pollen analysis and vegetational history of the Itasca region, Minnesota. *In* CUSHING, E.J. and WRIGHT, H.E., Jr., (eds.) *Quaternary Palaeoecology*, 219-36, New Haven.

McCALLIEN, W.J. 1937. Late-glacial and early Post-glacial Scotland. *Proc. Soc. Antiq. Scot.*, 71, 174-206.

McCANN, S.B. 1966. The Main Post-glacial Raised Shoreline of Western Scotland from the Firth of Lorne to Loch Broom. *Trans. Inst. Brit. Geogr.*, 39, 87-99.

McCANN, S.B. 1968. Raised rock platforms in the western Isles of Scotland. *In* BOWEN, E.G., CARTER, H. and TAYLOR, J.A. (eds.) *Geography at Aberystwyth*, 22-34, Cardiff.

McCAVE, I.N., CASTON, V.N.D. and FANNIN, N.G.T. 1977. The Quaternary of the North Sea. *In* SHOTTON, F.W. (ed.) *British Quaternary Studies - Recent Advances*, 187-204, Oxford.

MACFADYEN, W.A. 1942. A Post-glacial microfauna from Swansea Docks. *Geol. Mag.*, 79, 133-46.

MACFADYEN, W.A. 1955. Appendices 1,2. *In* GODWIN, H. Studies of the post-glacial history of British Vegetation XIII. The Meare Pool region of the Somerset Levels. *Phil. Trans. R. Soc. B*, 239, 185-90.

McINTYRE, A., KIPP, N.G., BÉ, A.H.W., CROWLEY, T., KELLOGG, T., GARDNER, J.V., PRELL, W. and RUDDIMAN, W.F. 1976. Glacial North Atlantic 18,000 years ago: a CLIMAP reconstruction. *In* CLINE, R.M. and HAYS, J.D. (eds.) Investigation of Late Quaternary Palaeoceanography and Palaeoclimatology (*Geol. Soc. Amer. Memoir*, 145), 43-76.

MACPHERSON, J.B. 1978. Pollen chronology of the Glen Roy-Loch Laggan proglacial lake drainage. *Scott. J. Geol.*, 14, 125-39.

MANGERUD, J. 1970. Late Weichselian vegetation and ice-front oscillations in the Bergen district, western Norway. *Norsk Geogr. Tidsskr.*, 24, 121-48.

MANGERUD, J. 1977. Late Weichselian marine sediments containing shells, foraminifera, and pollen, at Ågotnes, western Norway. *Norsk Geol. Tidsskr.*, 57, 23-54.

MANGERUD, J., ANDERSEN, S.T., BERGLUND, B.E. and DONNER, J.J. 1974. Quaternary stratigraphy of Norden, a proposal for terminology and classification. *Boreas*, 3, 109-27.

MANGERUD, J. and BERGLUND, B.E. 1978. The subdivision of the Quaternary of Norden: a discussion. *Boreas*, 7, 179-81.

MANGERUD, J. and GULLICKSEN, S. 1975. Apparent radiocarbon age of recent marine shells from Norway, Svalbard and Ellesmere Island. *Quat. Res.*, 5, 263-73.

MANGERUD, J., LARSEN, E., LONGVA, O. and SONSTEGAARD, E. 1979. Glacial history of western Norway 15,000 - 10,000 BP. *Boreas*, 8, 179-87.

MANLEY, G. 1959. The late-glacial climate of North-West England. *Liverp. and Manch. Geol. J.*, 2, 188-215.

MARTHINUSSEN, M. 1960. Coast and fjord areas of Finmark. *In* HOLTEDAHL, O. (ed.) *Geology of Norway*, 416-209, (*Norges Geol. Unders.*)

MENENDEZ AMOR, J. and FLORSCHUTZ, F. 1961. Resultado de analisis palinologico de algunas series de muestras de turba, arcilla y ostros sedimentos recogidos en los alrededores de I. Puebla de Sanabria (Zamora) II Buelna (Asturias). *Estudios Geol.*, 17, 83-99.

MENKE, B. 1968. Das Spätglazial von Glusing. *Eiszeit. u. Gegenw.*, 19, 73-84.

MERCER, J.H. 1969. The Alleröd oscillation: a European climatic anomaly? *Arctic and Alpine Res.*, 1, 227-34.

MERCER, J.H. 1972. The lower boundary of the Holocene. *Quat. Res.*, 2, 15-24.

MICHELSEN, O. 1967. Foraminifera of the Late-Quaternary deposits Laesø. *Meddr. dansk geol. Foren.*, 17, 205-63.

MILES, G.A. 1977. Planktonic foraminifera from Leg 37, deep sea drilling project. *In* AUMENTO, F., MELSON, W.C., *et al.*, (eds.) *Initial Reports of Deep-Sea Drilling Project, 1977*, vol. 37, 929-61.

MITCHELL, G.F. 1977. Periglacial Ireland. *Phil. Trans. R. Soc. Lond. B*, 280, 199-209.

MITCHELL, G.F., PENNY, L.F., SHOTTON, F.W. and WEST, R.G. 1973. A correlation of Quaternary deposits in the British Isles. *Geol. Soc. Lond. Spec. Rep.*, 4, 96 pp.

MOIGN, A. 1973. Strandflats immergés et emergés du Spitsberg Central et Nord-Occidental. Unpub. PhD Thesis, Univ. of Bretagne Occidentale.

MOIGN, A. 1974. Géomorphologie du strandflat au Svalbard; problèmes (âge, origine, processus) méthodes de travail. *Inter-Nord*, 13-14, 57-72.

MOIROUD, A. and GONNET, J.F. 1977. *Jardins de Glaciers*. Grenoble.

MOORE, P.D. 1970. Studies in the vegetational history of mid-Wales. II. The Late-glacial period in Cardiganshire. *New Phytol.*, 71, 947-59.

MOORE, P.D. and WEBB, J.A. 1978. *An Illustrated Guide to Pollen Analysis*. London.

MÖRNER, N.-A. (ed.) 1976. The Pleistocene/Holocene boundary. A proposed boundary-stratotype in Gothenburg, Sweden. *Boreas*, 5, 193-275.

MORRISON, R.B. 1969. The Pleistocene-Holocene Boundary: an evaluation of the various criteria used for determining it on a provincial basis, and suggestions for establishing it world-wide. *Geol. en Mijnb.*, 48, 363-71.

MOVIUS, H.L. 1942. *The Irish Stone Age*. Cambridge.

MÜLLER, Hanna. 1965. Vorkommen spätglazialer Tuffe in Nordostdeutschland. *Geologie*, 14, 1118-23.

MÜLLER, Helmut. 1953. Zur spät- und nacheiszeitlichen Vegetationsgeschichte des mitteldeutschen Trockengebietes. *Nova Acta Leopoldina*, 16, N.S., 1-67.

MÜLLER, H.J. 1972. Pollenanalytische Untersuchungen zum Eisrückzug und zur Vegetationsgeschichte im Vorderrhein- und Lukmaniergebiet. *Flora*, 161, 333-82.

MURRAY, J.W. 1973. *Distribution and Ecology of Living Benthic Foraminiferids*. London.

MURRAY, J.W. and HAWKINS, A.B. 1976. Sediment transport in the Severn Estuary during the past 8,000 - 9,000 years. *Q. J. Geol. Soc. Lond.*, 132, 385-98.

NAGY, J. 1965. Foraminifera in some bottom samples from shallow waters in Vestspitsbergen. *Norsk Polarinst. Årbok*, 1963, 109-28.

NANSEN, F. 1922. The strandflat and isostasy. *Videnskapselkapets Skrifter 1, Math.- Naturw. Kl., 1921; 11, Kristiana.*

NEALE, J.W. 1964. Some factors influencing the distribution of Recent British Ostracoda. *Pubbl. staz. zool. Napoli 33 suppl.*, 247-307.

NEWEY, W.W. 1970. Pollen analysis of Late Weichselian deposits at Corstorphine, Edinburgh. *New Phytol.*, 69, 1167-77.

NIEMELÄ, J. 1971. Die quartäre Stratigraphie von Tonablagerungen und der Rückzug des Inlandeises zwischen Helsinki und Hämeenlinna in Südfinnland. *Geol. Surv. Finland Bull.*, 253, 79 pp.

NIKLEWSKI, J. and van ZEIST, W. 1970. A late Quaternary pollen diagram from Northwestern Syria. *Acta Bot. Neerl.*, 19, 737-54.

NILSSON, T. 1961. Ein neues Standardpollendiagramm aus Bjärsjöhnolmssjön in Schonen. *Lunds Univ. Arsskr.*, N.F. 2:56, 34 pp.

NILSSON, T. 1965. The Pleistocene-Holocene boundary and the subdivision of the late Quaternary in southern Sweden. *INQUA, Rep. VIth Congr., Warsaw 1961*, 1, 479-94.

NYDAL, R., GULLIKSEN, S. and LØVSETH, K. 1972. Trondheim natural radiocarbon measurements VI. *Radiocarbon*, 14, 418-51.

OELE, E. 1968. The Quaternary Geology of the Dutch part of the North Sea, north of the Frisian Islands. *Geol. en Mijnb.*, 48, 467-80.

OELE, E. 1971. The Quaternary Geology of the southern area of the Dutch part of the North Sea. *Geol. en Mijnb.*, 50, 461-74.

OESCHGER, H., RIESEN, T. and LERMAN, J.C. 1970. Bern radiocarbon dates VII. *Radiocarbon*, 12, 358.

OESCHGER, H., STAUFFER, B., BUCHER, P. and MOELL, M. 1976. Extraction of trace components from large quantities of ice in bore holes. *J. Glaciol.*, 17, 117-28.

OGDEN, J.G. 1965. Radiocarbon determinations of sedimentation rates from hard and soft-water lakes in northeastern North America. *In* CUSHING, E.J. and WRIGHT, H.E. (eds.) *Quaternary Palaeoecology*, 175-83, New Haven.

OLDFIELD, F. 1960. Studies in the Post-glacial history of British vegetation: Lowland Lonsdale. *New Phytol.*, 59, 192-217.

OLSSON, I.U. 1972. The pretreatment of samples and the interpretation of the results of C14 determinations. *Acta Univ. Oul. A 3 Geol.*, 1, 9-37.

OLSSON, I.U. 1974. Some problems in connection with the evaluation of C14 dates. *Geol. Fören. Stockh. Förh.*, 96, 311-20.

OSBORNE, P.J. 1972. Insect faunas of Late Devensian and Flandrian age from Church Stretton, Shropshire. *Phil. Trans. R. Soc. Lond. B*, 263, 327-67.

OSTREM, G. 1965. Problems of dating ice-cored moraines. *Geogr. Annlr.*, 47A, 1-37.

O'SULLIVAN, P.E. 1974. Two Flandrian pollen diagrams from the East-central Highlands of Scotland. *Pollen Spores*, 16, 33-57.

O'SULLIVAN, P.E. 1975. Early and Middle-Flandrian pollen zonation in the Eastern Highlands of Scotland. *Boreas*, 4, 197-207.

O'SULLIVAN, P.E. 1976. Pollen analysis and radiocarbon dating of a core from Loch Pityoulish, Eastern Highlands of Scotland. *J. Biogeogr.*, 3, 293-302.

PATERSON, I.B. 1974. The supposed Perth Readvance in the Perth District. *Scott. J. Geol.*, 10, 53-66.

PATERSON, W.S.B. 1969. *The Physics of Glaciers.* Oxford.

PEACOCK. J.D. 1974. Borehole evidence for late- and postglacial events in the Cromarty Firth, Scotland. *Bull. Geol. Surv. Gt. Br.*, 48, 55-67.

PEACOCK, J.D. 1975a. Scottish late and post-glacial marine deposits. *In* GEMMELL, A.M.D. (ed.) *Quaternary Studies in North East Scotland*, 45-48, Aberdeen.

PEACOCK. J.D. 1975b Quaternary of Scotland - discussion. *Scott. J. Geol.*, 11, 174-5.

PEACOCK, J.D., GRAHAM, D.K., ROBINSON, J.E. and WILKINSON, I. 1977. Evolution and chronology of Lateglacial marine environments at Lochgilphead, Scotland. *In* GRAY, J.M. and LOWE, J.J. (eds.) *Studies in the Scottish Lateglacial Environment*, 89-100, Oxford.

PEACOCK, J.D., GRAHAM, D.K. and WILKINSON, I.P. 1978. Late-glacial and Post-glacial marine environments at Ardyne, Scotland, and their significance in the interpretation of the history of the Clyde sea Area. *Rep. Inst. Geol. Sci.*, No. 78/17, 1-24.

PEARSON, R.G. 1960. The coleoptera from some Late-Quaternary deposits and their significance for zoogeography. Unpub. Dissertation, Univ. of Cambridge.

PEARSON, R.G. 1962. Coleoptera from a Late-glacial deposit at St. Bees, West Cumberland. *J. An. Ecol.*, 31, 129-50.

PEGLAR, S. 1979. A radiocarbon-dated pollen diagram from Loch of Winless, Caithness, north-east Scotland. *New Phytol.*, 82, 245-63.

PENNINGTON, W. 1975a. A chronostratigraphic comparison of Late-Weichselian and Late-Devensian subdivisions, illustrated by two radiocarbon-dated profiles from western Britain. *Boreas*, 4, 157-71.

PENNINGTON, W. 1975b. An application of Principal Components Analysis to the zonation of two Late-Devensian profiles: Section II: Interpretation of the numerical analyses in terms of Late-Devensian (Late-Weichselian) environmental history. *New Phytol.*, 75, 441-53.

PENNINGTON, W. 1977a. The Late Devensian flora and vegetation of Britain. *Phil. Trans. R. Soc. Lond. B.* 280, 247-71.

PENNINGTON, W. 1977b. Lake sediments and the Lateglacial Environment in Northern Scotland. *In* GRAY, J.M. and LOWE, J.J. (eds.) *Studies in the Scottish Lateglacial Environment*, 119-42, Oxford.

PENNINGTON, W. 1978. Quaternary geology. *In* MOSELEY, F. (ed.) *Geology of the Lake District* (*Yorks. Geol. Soc.*, Occ. Publ. 3), 207-25, Leeds.

PENNINGTON, W. 1979. Comparisons between the Late-Devensian vegetation of Highland Britain and the present vegetation of West Greenland. Paper presented to Quaternary Research Association symposium, London, January 1979.

PENNINGTON, W. and BONNY, A.P. 1970. Absolute pollen diagram from the British Late Glacial. *Nature*, 226, 871-3.

PENNINGTON, W., HAWORTH, E.Y., BONNY, A.P. and LISHMAN, J.P. 1972. Lake sediments in Northern Scotland. *Phil. Trans. R. Soc. Lond. B*, 264, 191-294.

PENNINGTON, W. and SACKIN, M.J. 1975. An application of Principal Components Analysis to the zonation of two Late-Devensian profiles. I. Numerical analysis. *New Phytol.*, 75, 419-40.

PESCHKE, P. 1977. Zur Vegetations- und Besiedelungsgeschichte des Waldviertels (Niederösterreich). *Mitt. d. Komm. f. Quartärforschung d. Östereichischen Akad. d. Wissenschaften*, 2, 1-84.

PILCHER, J.R. 1973. Tree-ring research in Ireland. *Tree-Ring Bulletin*, 33, 1-28.

PORTER, S.C. and CARSON, R.J. 1971. Problems of interpreting radiocarbon dates from dead-ice terrain with an example from the Puget Lowland of Washington. *Quat. Res.*, 1, 410-14.

POST, A. 1976. The tilted forest: glaciological-geological implications of vegetated Neoglacial ice at Lituya Bay, Alaska. *Quat. Res.*, 6, 111-18.

PRICE, P.W. 1975. *Insect Ecology*. New York.

RALSKA-JASIEWICZOWA, M. 1966. Bottom sediments of the Mikolajki Lake (Masurian Lake District) in the light of Palaeobotanical Investigations. *Acta Palaeobot.*, 7, 3-118.

RALSKA-JASIEWICZOWA, M. 1972. The forests of the Polish Carpathians in the Late Glacial and Holocene. *Stud. Geomorphol. Carpatho-Baltica*, 6, 5-19.

RAUSCH, K.-A. 1975. Untersuchungen zur spät- und nacheiszeitlichen Vegetationsgeschichte im Gebiet des ehemaligen Inn-Chiemseegletschers. *Flora*, 164, 232-82.

REILLE, M. 1975. Contribution pollenanalytique a l'histoire tardiglaciaire et Holocène de la végétation de la Montagne Corse. Unpub. Thesis, Univ. of Aix-Marseille.

RICHMOND, G.M. (ed.) 1959. American Commission on Stratigraphic Nomenclature. Report 6: Committee on Pleistocene, Application of Stratigraphic Classification and Nomenclature to the Quaternary. *Bull. Am. Ass. Petrol. Geol.*, 43, 663-73.

RIEGER, S. 1974. Arctic soils. *In* IVES, J.D. and BARRY, R.G. (eds.) *Arctic and Alpine Environments*, 749-69, London.

ROBERTSON, D. 1875. Notes on the recent ostracoda and foraminifera of the Firth of Clyde, with some remarks on the distribution of mollusca. *Trans. Geol. Soc. Glasgow*, 5, 112-53.

ROBINSON, G., PETERSON, J.A. and ANDERSON, P.K. 1971. Trend surface analysis of corrie altitudes in Scotland. *Scott. Geogr. Mag.*, 87, 142-6.

ROBINSON, E. 1978. The Pleistocene. *In* BATE, R.H. and ROBINSON, E. (eds.) *A Stratigraphical Index of British Ostracoda*. Liverpool.

ROKOENGEN, K., BELL, G., BUGGE, T., DEKKO, T., GUNNLEIKSRUD, T., LIEN, R.L., LØFALDLI, M., and VIGRAN, J.O. 1977. Provetaking av fjellgrunn og løsmasser utenfor deler av Nord-Norge i 1976. *Cont. Shelf Inst., Publ.*, 91, 1-65 (Trondheim).

ROSE, J., TURNER, C., COOPE, G.R. and BRYAN, M.D. 1979. River channel changes in a low relief catchment over the last 13,000 years. *In* CULLINGFORD, R.A., DAVIDSON, D.A. and LEWIN, J. (eds.) *Timescales in Geomorphology* (in the press), New York.

ROWLANDS, P.H. and SHOTTON, F.W. 1971. Pleistocene deposits of Church Stretton (Shropshire) and its neighbourhood. *Q.J. Geol. Soc. Lond.*, 127, 599-622.

RUDDIMAN, W.F. and McINTYRE, A. 1973. Time-transgressive deglacial retreat of polar waters from the North Atlantic. *Quat. Res.*, 3, 117-30.

RUDDIMAN, W.F., SANCETTA, C.D. and McINTYRE, A. 1977. Glacial/interglacial response rate of subpolar North Atlantic water to climatic change: the record in ocean sediments. *Phil. Trans. R. Soc. Lond. B*, 280, 119-42.

RYBNÍČKOVÁ, E. and RYBNÍČEK, K. 1972. Erste Ergebnisse paläeogeobotanischer Untersuchungen des Moores bei Vracov, Südmähren. *Folia Geobot. Phytotax., Praha*, 7, 285-308.

RYMER, L. 1977. A Late-glacial and early Post-glacial pollen diagram from Drimnagall, North Knapdale, Argyllshire. *New Phytol.*, 79, 211-221.

SCHMEIDL, H. 1971. Ein Beitrag zur spätglazialen Vegetations- und Waldentwicklung im westlichen Salzachgletschergebiet. *Eiszeit. u. Gegenw.*, 22, 110-16.

SCHNEIDER, R.E. 1978. Pollenanalytische Untersuchungen zur Kenntnis der spät- und postglazialen Vegetationsgeschichte am Südrand der Alpen zwischen

Turin und Varese (Italien). *Bot. Jahrb. Syst.*, 100, 26-109.

SCHREVE-BRINKMAN, E.J. 1978. A palynological study of the Upper Quaternary sequence in the El Abra corridor and rock shelters (Colombia). *Palaeogeogr. Palaeoclimatol. Palaeoecol.*, 25, 1-109.

SEDDON, B. 1962. Late-glacial deposits at Llyn Dwythwch and Nant Ffrancon, Caernarvonshire. *Phil. Trans. R. Soc. Lond. B*, 244, 459-81.

SERCELJ, A. 1971. Postglazialni razvoj gorskih gozdiv v severozahodni Jugoslaviji (Die postglaziale Entwicklung der Gebirgswälder im nordwestlichen Jugoslawien). *Razpr. 4. r. Slov. akad. znan. umetn.*, 9, 267-93.

SHACKLETON, N.J. 1977. Oxygen isotope stratigraphy of the Middle Pleistocene. *In* SHOTTON, F.W. (ed.) *British Quaternary Studies - Recent Advances*, 1-16, Oxford.

SHOTTON, F.W. 1967. The problems and contributions of methods of absolute dating within the Pleistocene period. *Q. J. Geol. Soc. Lond.*, 122, 357-83.

SHOTTON, F.W. 1972. An example of hard-water error in radiocarbon dating of vegetable matter. *Nature*, 240, 460-1.

SIMPKINS, K.S. 1974. The Late-glacial deposits at Glanllynau, Caernarvonshire. *New Phytol.*, 73, 605-18.

SINGH, G. 1970. Late-glacial vegetational history of Lecale, Co. Down. *Proc. R. Irish Acad.*, B, 69, 189-216.

SISSONS, J.B. 1967. *The Evolution of Scotland's Scenery*. Edinburgh.

SISSONS, J.B. 1972. The last glaciers in part of the south-east Grampians. *Scott. Geogr. Mag.*, 88, 168-81.

SISSONS, J.B. 1974a. A late-glacial ice cap in the central Grampians, Scotland. *Trans. Inst. Brit. Geogr.*, 62, 95-114.

SISSONS, J.B. 1974b. The Quaternary in Scotland: a review. *Scott. J. Geol.*, 10, 311-37.

SISSONS, J.B. 1974c. Lateglacial marine erosion in Scotland. *Boreas*, 3, 41-48.

SISSONS, J.B. 1976a. A remarkable protalus rampart in Wester Ross. *Scott. Geogr. Mag.*, 92, 182-90.

SISSONS, J.B. 1976b. *The Geomorphology of the British Isles: Scotland*. London.

SISSONS, J.B. 1977a. The Loch Lomond Readvance in southern Skye and some palaeoclimatic implications. *Scott. J. Geol.*, 13, 23-36.

SISSONS, J.B. 1977b. The Loch Lomond Readvance in the Northern mainland of Scotland. *In* GRAY, J.M. and LOWE, J.J. (eds.) *Studies in the Scottish Lateglacial Environment*, 45-59, Oxford.

SISSONS, J.B. 1977c. Former ice-dammed lakes in Glen Moriston, Inverness-shire and their significance in upland Britain. *Trans. Inst. Brit. Geogr.*, N.S. 2, 224-42.

SISSONS, J.B. 1978. The parallel roads of Glen Roy and adjacent glens, Scotland. *Boreas*, 7, 229-44.

SISSONS, J.B. 1979a. The Loch Lomond Advance in the Cairngorm Mountains. *Scott. Geogr. Mag.* (in the press).

SISSONS, J.B. 1979b. The limit of the Loch Lomond Advance in Glen Roy and vicinity. *Scott. J. Geol.*, 15, 31-42.

SISSONS, J.B. 1979c. The Loch Lomond Stadial in the British Isles. *Nature*, 280, 199-203.

SISSONS, J.B. 1979d. Palaeoclimatic inferences from former glaciers in Scotland and the Lake District. *Nature*, 278, 518-21.

SISSONS, J.B. (in the press). The Loch Lomond Advance in the Lake District.

SISSONS, J.B. and GRANT, A.J.H. 1972. The last glaciers in the Lochnagar area, Aberdeenshire. *Scott. J. Geol.*, 8, 85-93.

SISSONS, J.B. and SUTHERLAND, D.G. 1976. Climatic inferences from former glaciers in the south-east Grampian Highlands, Scotland. *J. Glaciol.*, 17, 325-46.

SISSONS, J.B. and WALKER, M.J.C. 1974. Late glacial site in the central Grampian Highlands. *Nature*, 249, 822-4.

SMITH, B. 1912. The glaciation of the Black Combe district. *Q. J. Geol. Soc. Lond.*, 68, 402-48.

SMITH, J. 1839. On the last changes in the relative levels of land and sea in the west of Scotland, the British Isles. *Trans. Wernerian Soc.*, VIII, 49-64.

SOLLID, J.L., ANDERSEN, S., HAMRE, N., KJELDSEN, O., SALVIGSEN, O., STURØD, S., TVEITA, T. and WILHELMSEN, A. 1973. Deglaciation of Finmark, North Norway. *Norsk geol. Tidsskr.*, 27, 233-325.

SOLLID, J.L. and SØRBEL, L. 1975. Younger Dryas ice-marginal deposits in Trondelag, central Norway. *Norsk geogr. Tidsskr.*, 29, 1-9.

SOLLID, J.L. and SØRBEL, L. 1979. Deglaciation of western Central Norway. *Boreas*, 8, 233-39.

SØRENSEN, R. 1979. Late Weichselian deglaciation in the Oslofjord area, south Norway. *Boreas*, 8, 241-46.

SOUCHEZ, R. and LORRAIN, R.D. 1975. Chemical sorting effect at the base of an alpine glacier. *J. Glaciol.*, 14, 261-5.

SOUCHEZ, R. and LORRAIN, R.D. 1978. Origin of the basal ice layer from Alpine glaciers indicated by its chemistry. *J. Glaciol.*, 20, 319-28.

SOUCHEZ, R., LORRAIN, R.D. and LEMMENS, M.M. 1973. Refreezing of interstitial water in a subglacial cavity of an Alpine glacier as indicated by the chemical composition of ice. *J. Glaciol.*, 12, 453-60.

STAUFFER, B. and BERNER, W. 1978. CO_2 in natural ice. *J. Glaciol.*, 21, 291-9.

STEERS, J.A. 1952. The coastline of Scotland. *Geogr. J.*, 118, 180-90.

STEPHENS, N. 1957. Some observations on the "Interglacial" Platform and the Early Post-glacial Raised Beach on the east coast of Ireland. *Proc. R. Irish Acad.*, 58, B6, 129-49.

STOCKMARR, J. 1971. Tablets with spores used in absolute pollen analysis. *Pollen Spores*, 13, 615-21.

STOCKMARR, J. 1975. Biostratigraphic studies in Late Weichselian sediments near Böllingsö. *Danm. Geol. Unders., Årbog.* 1974, 71-89.

STRAHAN, A., FLETT, J.S. and DINHAM, C.H. 1917. Special reports on the mineral resources of Great Britain, volume V. *Mem. Geol. Surv. Gt. Br.*

STUIVER, M. 1975. Climate versus changes in ^{13}C content of the organic component of lake sediments during the late Quaternary. *Quat. Res.*, 5, 251-62.

STUIVER, M. and POLACH, H.A. 1977. Reporting of C14 data. *Radiocarbon*, 19, 355-63.

SWITSUR, V.R. and WEST, R.G. 1975. University of Cambridge natural radiocarbon measurements XIV. *Radiocarbon*, 17, 301-12.

SYNGE, F.M. 1966. The relationship of the raised strandlines and main end moraines on the Isle of Mull, and in the district of Lorn, Scotland. *Proc. Geol. Assoc.*, 77, 315-28.

TAUBER, H. 1960. Copenhagen radiocarbon dates IV. *Radiocarbon*, 2, 12-25.

TAUBER, H. 1970. The Scandinavian varve chronology and C-14 dating. *In* OLSSON, I.U. (ed.) *Radiocarbon Variations and Absolute Chronology* (Proc. 12th. Nobel Symposium, Uppsala 1969), 173-95, Stockholm.

THIELE, H-U. 1977. *Carabid Beetles in their Environments*. New York.

THOMPSON, K.S.R. 1972. The last glaciers in western Perthshire. Unpub. PhD Thesis, Univ. of Edinburgh.

THOMSON, M.E. 1977. IGS studies of the geology of the Firth of Forth, and its approaches. *Rep. Inst. Geol. Sci.* No. 77/17, 1-56.

THORP, P.W. 1979. The Loch Lomond Readvance in the Glen Nevis and Loch Leven areas of the Western Grampians. Unpub. MSc Thesis, City of London Poly. and Poly. of North London.

TROTTER, F.M. and HOLLINGWORTH, S.E. 1932. The glacial sequence in the north of England. *Geol. Mag.*, 69, 374-80.

TUTIN, T.G., HEYWOOD, V.H., BURGES, N.A., MOORE, D.M., VALENTINE, D.H., WALTERS, S.M., and WEBB, D.A. 1972. *Flora Europaea. Vol. 3 Diapensiaceae to Myoporaceae.* Cambridge.

TUTIN, T.G., HEYWOOD, V.H., BURGES, N.A., MOORE, D.M., VALENTINE, D.H., WALTERS, S.M., and WEBB, D.A. 1976. *Flora Europaea. Vol.4 Plantaginaceae to Compositae (and Rubiaceae).* Cambridge.

USINGER, H. 1975. Pollenanalytische und stratigraphische Untersuchungen an zwei Spätglazial-Vorkommen in Schleswig-Holstein. *Mitt. d. Arbeitsgemein-*

schaft Geobotanik in Schleswig-Holstein u. Hamburg, 25, 1-183.

VASARI, Y. 1977. Radiocarbon dating of the Lateglacial and early Flandrian vegetational succession in the Scottish Highlands and the Isle of Skye. *In* GRAY, J.M. and LOWE, J.J. (eds.) *Studies in the Scottish Lateglacial Environment*, 143-62, Oxford.

VASARI, Y. and VASARI, A. 1968. Late- and Post-glacial macrophytic vegetation in the lochs of northern Scotland. *Acta. bot. Fenn.*, 80, 120 pp.

VERNEKAR, A.D. 1971. Long-period global variations of incoming solar radiation. *Meteorol. Monographs*, 12, No.34.

VITA-FINZI, C. 1973. *Recent Earth History*. London.

VOGEL, J.C. 1970. C14 trends before 6,000 BP. *In* OLSSON, I.U. (ed.) *Radiocarbon Variations and Absolute Chronology* (Proc. 12th. Nobel Symp., Uppsala 1969), 313-25, Stockholm.

VOGT, T. 1918. Om recente og gamle strandlinjer i fast fjeld. *Norsk Geol. Tidsskr.*, III, 8, Kristiana.

VORREN, K.D. 1978. Late and Middle Weichselian stratigraphy of Andøya, north Norway. *Boreas*, 7, 19-38.

VORREN, T.O. and ELVSBORG, A. 1979. Late Weichselian deglaciation and palaeoenvironment of the shelf and coastal areas of Troms, north Norway - a review. *Boreas*, 8, 247-53.

VORREN, T.O., STRASS, I.E. and LIND-HANSEN, O.W. 1978. Late Quaternary sediments and stratigraphy on the continental shelf off Troms and west Finnmark, northern Norway. *Quat. Res.*, 10, 340-65.

WALKER, D. 1956. A Late-glacial deposit at St. Bees, Cumberland. *Q. J. Geol. Soc. Lond.*, 112, 93-101.

WALKER, D. and GODWIN, H. 1954. *Excavations at Starr Carr*. Cambridge.

WALKER, M.J.C. 1975a. Two Lateglacial pollen diagrams from the eastern Grampian Highlands, Scotland. *Pollen Spores*, 17, 67-92.

WALKER, M.J.C. 1975b. Lateglacial and early Postglacial environmental history of the central Grampian Highlands, Scotland. *J. Biogeogr.* 2, 265-84.

WALKER, M.J.C. 1977. Corrydon: a Lateglacial profile from Glenshee, South-east Grampian Highlands, Scotland. *Pollen Spores*, 19, 391-406.

WALKER, M.J.C. and LOWE, J.J. 1977. Postglacial environmental history of Rannoch Moor, Scotland. I. Three pollen diagrams from the Kingshouse area. *J. Biogeogr.*, 4, 333-51.

WALKER, M.J.C. and LOWE, J.J. 1979a. Postglacial environmental history of Rannoch Moor, Scotland. II. Pollen analyses and radiocarbon dates from the Rannoch Station and Corrour areas. *J. Biogeogr.*, 6, (in the press).

WALKER, M.J.C. and LOWE, J.J. 1979b. Pollen analyses, radiocarbon dates and the deglaciation of Rannoch Moor, Scotland, following the Loch Lomond Advance. *In* CULLINGFORD, R.A., DAVIDSON, D.A. and LEWIN, J. (eds.), *Timescales in Geomorphology*, (in the press), New York.

WALTON, W.R. 1964. Recent foraminiferal ecology and palaeoecology. *In* IMBRIE, J. and NEWELL, N.D. (eds.) *Approaches to Palaeoecology*, 151-237, New York.

WASYLIKOWA, K. 1964a. Roślinośc i Klimat Póżnego Glacjalú w Sródkowej Polsce no podstawie Badań w Witowie kolo Leczycy. *Biul. Peryglac.*, 13, 261-417.

WASYLIKOWA, K. 1964b. Pollen analysis of the Late Glacial sediments in Witow near Leczyca, Middle Poland. *INQUA Rep. VIth Congr., Warsaw 1961*, 2, 497-502.

WATTS, W.A. 1973. Rates of change and stability in vegetation in the perspective of long periods of time. *In* WEST, R.G. and BIRKS, H.J.B. (eds.), *Quaternary Plant Ecology*, 195-206, Oxford.

WATTS, W.A. 1977. The Late Devensian vegetation of Ireland. *Phil. Trans. R. Soc. Lond. B*, 280, 273-93.

WATTS, W.A. 1979. Late Quaternary vegetation of Central Appalachia and the New Jersey Coastal Plain. *Ecol. Monographs*, (in the press).

WEBB, J.A. 1977. Studies on the Late Devensian vegetation of the Whitlaw Mosses, South-east Scotland. Unpub. PhD Thesis, Univ. of London.

WEBB, T. III and BRYSON, R.A. 1972. Late and Post Glacial climatic change in the Northern Midwest, USA: Quantitative estimates derived from fossil pollen spectra by multivariate statistical analysis. *Quat. Res.*, 2, 70-115.

WEGMÜLLER, S. 1966. *Über die spät- und postglaziale Vegetationsgeschichte des südwestlichen Jura*. Bern.

WEGMÜLLER, S. 1977. *Pollenanalytische Untersuchungen zur spät- und postglazialen Vegetationsgeschichte der französischen Alpen (Dauphiné)*. Bern.

WEGMÜLLER, S. and WELTEN, M. 1973. Spätglaziale Bimstufflagen des Laacher Vulkanismus im Gebiet der westlichen Schweiz und der Dauphiné (F). *Eclog. Geol. Helv.*, 66, 533-41.

WEISS, R.F., BUCHER, P., OESCHGER, H. and CRAIG, H. 1972. Compositional variations of gases in temperate glaciers. *Earth and Planetary Sci. Letters*, 16, 178-84.

WENTWORTH, C.K. 1938. Marine bench forming processes: water level weathering. *J. Geomorph.*, 1, 6-32.

WEST, R.G. 1977. *Pleistocene Geology and Biology*. London.

WHATLEY, R.C. 1976. Association between podocopid Ostracoda and some animal substrates. *Abh. Verh. Naturwiss. Verein Hamburg*, N.F. 18/19 suppl., 191-200.

WIJMSTRA, T.A. 1969. Palynology of the first 30 metres of a 120 m deep section in northern Greece. *Acta. Bot. Neerl.*, 18, 511-27.

WILLIAMS, R.B.G. 1975. The British climate during the last glaciation; an interpretation based on periglacial phenomena. *In* WRIGHT, A.E. and MOSELEY, F.M. (eds.) *Ice Ages: Ancient and Modern* (Geol. J. Spec. Issue, 6), 95-117, Liverpool.

WILLIAMS, R.E.G. and JOHNSON, A.S. 1976. Birmingham University radiocarbon dates X. *Radiocarbon*, 18, 249-67.

WILLIAMS, W. 1976. The Flandrian vegetation of the Isle of Skye and the Morar Peninsula. Unpub. PhD Thesis, Univ. of Cambridge.

WORSLEY, P. 1977. Periglaciation. *In* SHOTTON, F.W. (ed.) *British Quaternary Studies: Recent Advances*, 205-19, Oxford.

WRIGHT, H.E. Jr. 1971. Late Quaternary vegetational history of North America. *In* TUREKIAN, K.K. (ed.) *The Late Cenzoic Glacial Ages*, 425-64, New Haven.

WRIGHT, H.E. Jr. 1977. Environmental change and the origin of agriculture in the Old and New Worlds. *In* REED, C.A. (ed.) *Origins of Agriculture*, 281-318, The Hague.

WRIGHT, W.B. 1928. The raised beaches of the British Isles. *First Rep. of Comm. on Pliocene and Pleistocene Terraces* (Int. Geogr. Union), 99-106.

WRIGHT, W.B. 1937. *The Quaternary Ice Age*. London.

YOUNG, J.A.T. 1978. The landforms of upper Speyside. *Scott. Geogr. Mag.*, 94, 76-94.

van ZEIST, W. 1967. Late Quaternary vegetational history of western Iran. *Rev. Palaeobotan. Palynol.*, 2, 301-11.

ZENKOVICH, V.P. 1967. *Processes of Coastal Development*. (STEERS, J.A., editor), 383-493, Edinburgh.

ZOLLER, H. 1960. Pollenanalytische Untersuchungen zur Vegetationsgeschichte der insubrischen Schweiz. *Denkschr. Schweiz Naturf. Ges.*, 88, 45-157.

ZOLLER, H. and KLEIBER, H. 1971a. Vegetationsgeschichtliche Untersuchungen in der montanen und subalpinen Stufe der Tessintäler. *Verh. Naturf. Ges. Basel*, 81, 90-154.

ZOLLER, H. and KLEIBER, H. 1971b. Überblick der spät- und postglazialen Vegetationsgeschichte in der Schweiz. *Boissiera*, 19, 113-28.

ADDITIONAL REFERENCE

REKSTAD, J. 1915. Om strandlinjer og strandlinjedannelse. *Norsk geol. Tidsskr.*, III, 8, Kristiana.

Index